SANITATION DETAILS

D1628533

Leslie Woolley

TD, FIPHE, Companion Chartered Institution
of Building Services, MRSH

Edited by Phil Stronach
Technical consultant Graham Fox

Taylor & Francis
Taylor & Francis Group

LONDON AND NEW YORK

© International Thomson Publishing Ltd

A 'Building Trades Journal' book

First published 1974
Reprinted 1977
Revised 1990
Reprinted 2003
By Taylor & Francis,
2 Park Square, Milton Park, Abingdon, Oxon, OX14 4RN

Transferred to Digital Printing 2005

ISBN 0 7198 2610 1

Contents

MAIN PURPOSES OF SANITARY APPLIANCES

Requirements
Appliances should be impervious, quiet in operation, easy to clean and maintain, and of a convenient shape, size and height. They should provide for the removal of contents quickly and completely; overflows should be adequate to deal with maximum inflow. Wastes and overflows should be screened against entry of anything that may cause obstruction.

Inspection and testing
All appliances, materials and workmanship to be carefully examined for defects upon completion. Comprehensive tests of all appliances should be made by simulating conditions of use. Also check overflows.

Selection factors
1. Cost
2. Hygiene
3. Maintenance
4. Durability
5. Appearance
6. Size and function
7. Weight (from a fixing point of view)
8. Thermal capacity
9. Speed to fill and empty
10. Noise
11. Resilience
12. Ease of installation

General
Pipes connected to appliances to be accessible for repair. In large buildings, a stopvalve should be fitted on all branch pipes. Precautions to prevent back flow should be observed.

Maintenance
Appliances need frequent cleaning to maintain them in good condition and appearance. If neglected, use solution of chloride of lime with hot water on ceramic ware to remove surface stains. Paraffin-moistened cloth and hot water may restore lustre.

1. Removal of human excreta

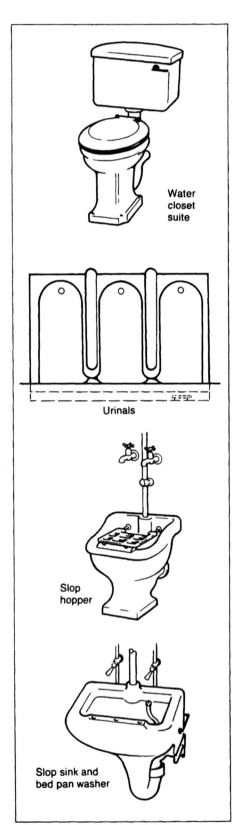

Water closet suite

Urinals

Slop hopper

Slop sink and bed pan washer

2. Personal ablution

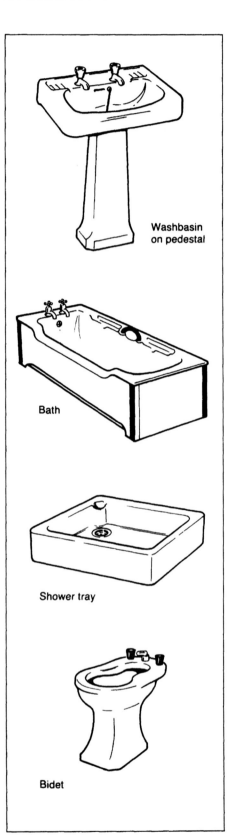

Washbasin on pedestal

Bath

Shower tray

Bidet

Sanitation: Appliances

3. Food preparation

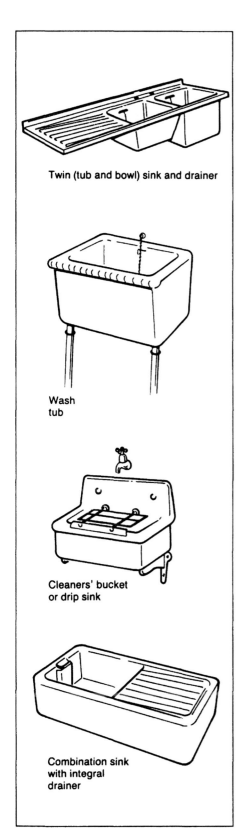

Twin (tub and bowl) sink and drainer

Wash
tub

Cleaners' bucket
or drip sink

Combination sink
with integral
drainer

4. Washing of utensils, clothes or floors

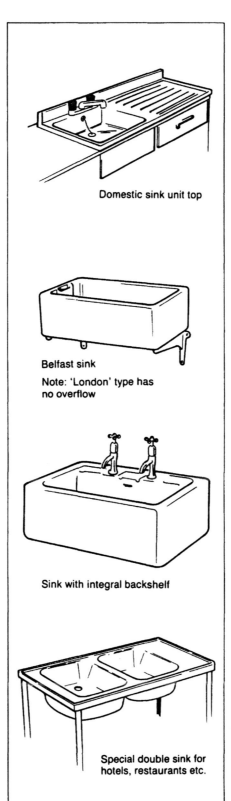

Domestic sink unit top

Belfast sink

Note: 'London' type has
no overflow

Sink with integral backshelf

Special double sink for
hotels, restaurants etc.

5. Special purposes in factories, schools, hospitals, dental surgeries etc.

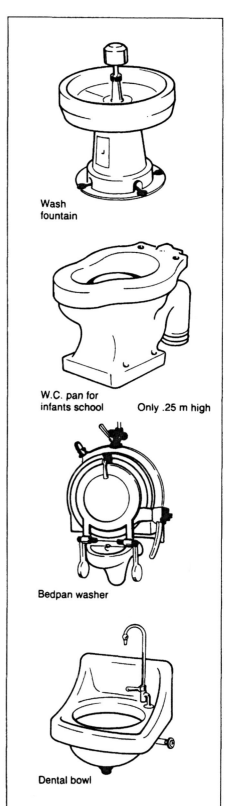

Wash
fountain

W.C. pan for
infants school   Only .25 m high

Bedpan washer

Dental bowl

Waste Appliances A Selection of Types

WASH BASINS

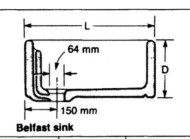

Plan

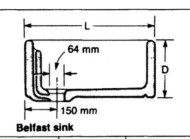

Bracket

785 mm

Floor level

A	635 mm	560 mm	510 mm
B	460 mm	405 mm	405 mm
C	480mm	405 mm	355 mm
D	585 mm	510 mm	460 mm

*Complies with BS.1188

180-200 mm

785 mm

Wash basin on a pedestal

330 or 395 mm

510 or 430 mm

540 or 460 mm

240 or 265 mm

Corner wash basin   cross section

SINKS

L

64 mm

D

150 mm

Belfast sink

Length	Width	Depth
915 mm	610 mm	305 mm*
915 mm	510 mm	255 mm
915 mm	460 mm	255 mm
760 mm	460 mm	255 mm*
685 mm	460 mm	255 mm
610 mm	460 mm	255 mm*
610 mm	405 mm	255 mm*
530 mm	405 mm	255 mm
760 mm	460 mm	200 mm
610 mm	460 mm	200 mm*
610 mm	405 mm	200 mm*
530 mm	405 mm	200 mm
460 mm	380 mm	200 mm*

*Sizes to BS.1206

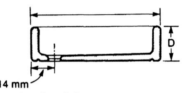

D

114 mm

London sink

Length	Width	Depth
610 mm	460 mm	255 mm
460 mm	380mm	200 mm

These sinks are glazed all round and are therefore reversible.

Sink with back shelf
(Fixing height — 0.9m: Top of front edge)

Plan

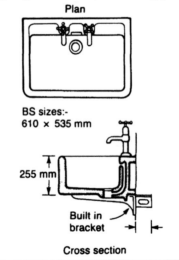

BS sizes:-
610 × 535 mm

255 mm

Built in bracket

Cross section

BIDETS

Rim

530 mm

380 mm

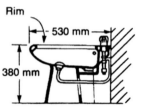

Hot to spray   Hot to rim   Cold to spray

170 mm

Spray

330 mm

Plan   Rear elevation

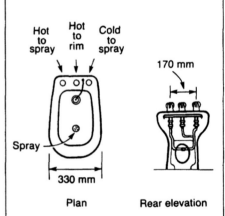

Notes. Bidet is French for 'little horse' — to be used by sitting astride the appliance — so designed that the unpleasant task of cleansing the excretory organs is achieved in a thorough and consistent manner — produces a pleasant feeling of complete personal cleanliness, with the minimum of inconvenience — able to solve the periodic problem of personal hygiene for the female sex — bidets have a secondary use as a footbath — rim can be warmed by the hot flush water — the water for the ascending spray is blended.

Sanitation: Appliances

SHOWER TRAYS

Shower and foot bath trays

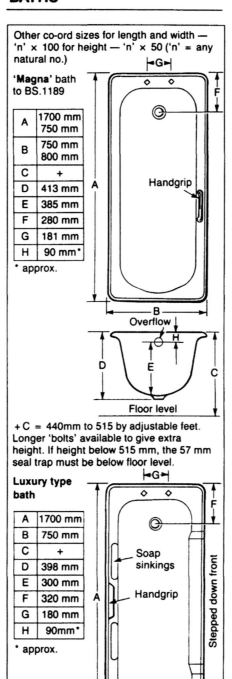

With overflow

915 × 915 × 180 mm
760 × 760 × 180 mm
610 × 610 × 180 mm

With overflow

915 × 915 × 180 mm
760 × 760 × 180 mm
610 × 610 × 180 mm

915 × 915 × 180 mm
760 × 760 × 180 mm
610 × 610 × 180 mm

915 × 915 × 180 mm
760 × 760 × 180 mm

Floor channel

BATHS

Other co-ord sizes for length and width — 'n' × 100 for height — 'n' × 50 ('n' = any natural no.)

'Magna' bath to BS.1189

A	1700 mm
	750 mm
B	750 mm
	800 mm
C	+
D	413 mm
E	385 mm
F	280 mm
G	181 mm
H	90 mm*

* approx.

Handgrip

Overflow

Floor level

+ C = 440mm to 515 by adjustable feet. Longer 'bolts' available to give extra height. If height below 515 mm, the 57 mm seal trap must be below floor level.

Luxury type bath

A	1700 mm
B	750 mm
C	+
D	398 mm
E	300 mm
F	320 mm
G	180 mm
H	90mm*

* approx.

Soap sinkings

Handgrip

Stepped down front

+ C = 470 mm to 540 mm by adjustable feet. Longer 'bolts' available to give extra height.

DRINKING FOUNTAINS

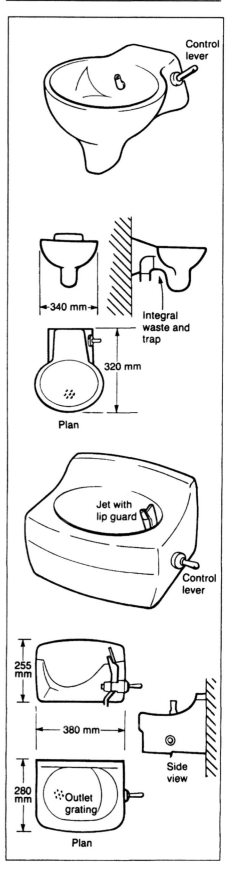

Control lever

340 mm

Integral waste and trap

320 mm

Plan

Jet with lip guard

Control lever

255 mm

380 mm

Side view

280 mm

Outlet grating

Plan

2

Soil Appliances (1) A Selection of W.C. Pans

WASHDOWN WATER CLOSET PAN TO BS.5503:1981

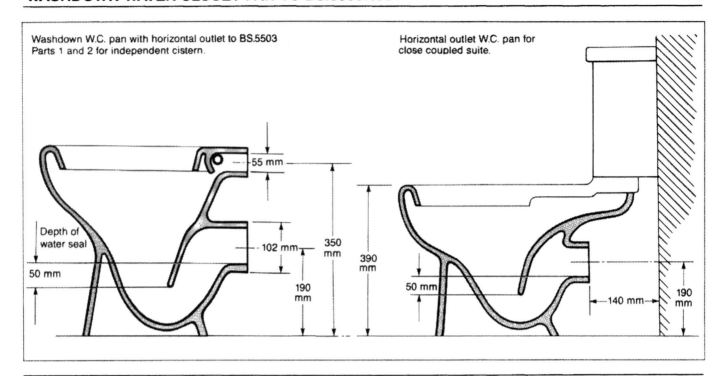

Washdown W.C. pan with horizontal outlet to BS.5503 Parts 1 and 2 for independent cistern.

Horizontal outlet W.C. pan for close coupled suite.

DEFINITIONS OF SOIL APPLIANCE

'A sanitary appliance for the collection and discharge of excretory matter' (C.P.305)

As used for interpretation of The Building Regulations — 'Soil appliance includes a water closet or urinal receptacle, bed-pan washer, bed-pan sink and slop sink'.

SIPHONIC TYPES

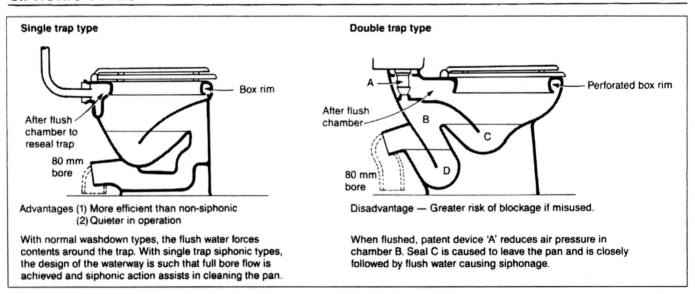

Single trap type

Advantages (1) More efficient than non-siphonic
(2) Quieter in operation

With normal washdown types, the flush water forces contents around the trap. With single trap siphonic types, the design of the waterway is such that full bore flow is achieved and siphonic action assists in cleaning the pan.

Double trap type

Disadvantage — Greater risk of blockage if misused.

When flushed, patent device 'A' reduces air pressure in chamber B. Seal C is caused to leave the pan and is closely followed by flush water causing siphonage.

Sanitation: Appliances

VARIOUS W.C. ARRANGEMENTS

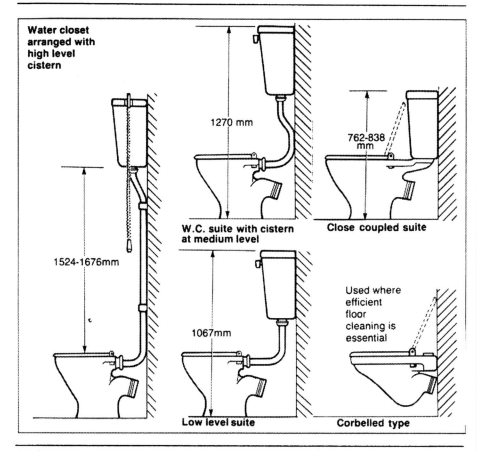

Water closet arranged with high level cistern

1270 mm

762-838 mm

1524-1676mm

1067mm

W.C. suite with cistern at medium level

Close coupled suite

Used where efficient floor cleaning is essential

Low level suite

Corbelled type

MISCELLANEOUS WATER CLOSET PANS

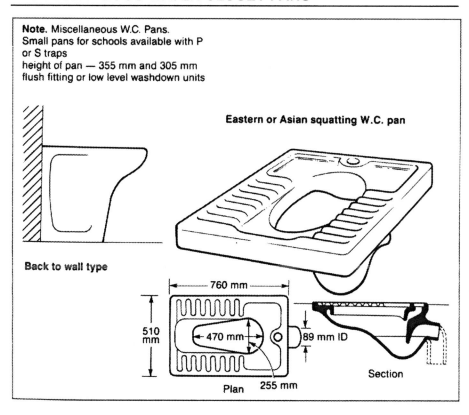

Note. Miscellaneous W.C. Pans.
Small pans for schools available with P or S traps
height of pan — 355 mm and 305 mm
flush fitting or low level washdown units

Eastern or Asian squatting W.C. pan

Back to wall type

760 mm

510 mm

470 mm

89 mm ID

Section

Plan 255 mm

W.C. SEATS

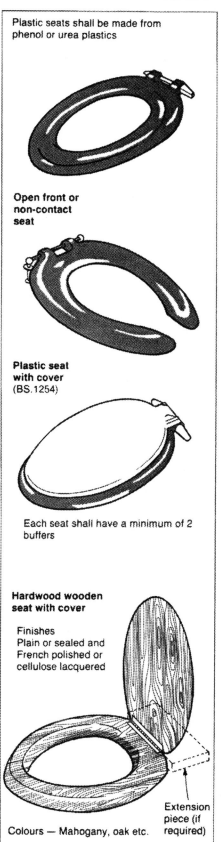

Plastic seats shall be made from phenol or urea plastics

Open front or non-contact seat

Plastic seat with cover
(BS.1254)

Each seat shall have a minimum of 2 buffers

Hardwood wooden seat with cover

Finishes
Plain or sealed and French polished or cellulose lacquered

Colours — Mahogany, oak etc.

Extension piece (if required)

3

TYPES OF URINAL

There are three main types of urinal:
(i) Made to measure slab urinals, with or without divisions;
(ii) Slab urinal ranges in modules, i.e. two persons to give persons; and
(iii) one-piece urinal available with concealed pipework for high risk areas.

The above units are cistern fed; which can be connected to a water saving device i.e. electronically operated solenoid valve, flow control device with illuminated manual on/off switch.

BS.6465 Part 1:1984 states that the provision of bowl type urinals should only be considered for areas of low risk where responsibility in use can be anticipated.

Slab and stall urinals should be flushed by means of a spreader or perforated sparge pipe which should cleanse the whole surface likely to be fouled.

TROUGHS

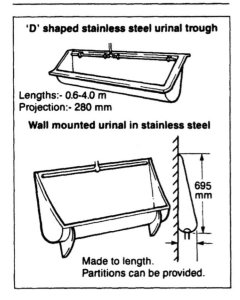

'D' shaped stainless steel urinal trough

Lengths:- 0.6-4.0 m
Projection:- 280 mm

Wall mounted urinal in stainless steel

695 mm

Made to length.
Partitions can be provided.

STALL URINAL · SLAB URINAL · ONE PIECE URINAL

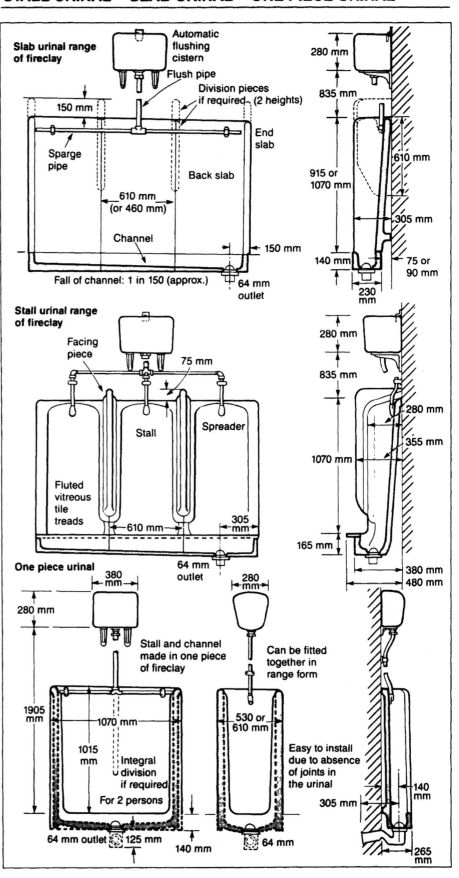

Slab urinal range of fireclay

Automatic flushing cistern
Flush pipe
Division pieces if required (2 heights)
150 mm
End slab
Sparge pipe
Back slab
610 mm (or 460 mm)
Channel
150 mm
Fall of channel: 1 in 150 (approx.)
64 mm outlet

280 mm
835 mm
915 or 1070 mm
610 mm
305 mm
140 mm
75 or 90 mm
230 mm

Stall urinal range of fireclay

Facing piece
75 mm
Stall
Spreader
Fluted vitreous tile treads
610 mm
305 mm
64 mm outlet

280 mm
835 mm
280 mm
355 mm
1070 mm
165 mm
380 mm
480 mm

One piece urinal

380 mm
280 mm
Stall and channel made in one piece of fireclay
1905 mm
1070 mm
1015 mm
Integral division if required
For 2 persons
64 mm outlet
125 mm
140 mm

280 mm
530 or 610 mm
Can be fitted together in range form
Easy to install due to absence of joints in the urinal
64 mm

140 mm
305 mm
265 mm

Sanitation: Appliances

BOWL URINALS

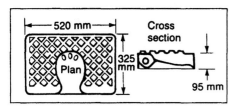

Side elevation | Front elevation

250 mm
300 mm
460 mm
100 mm

SQUATTING URINAL

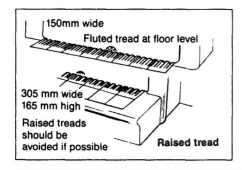

520 mm
Cross section
Plan
325 mm
95 mm

TREAD DETAILS

150mm wide
Fluted tread at floor level

305 mm wide
165 mm high

Raised treads should be avoided if possible

Raised tread

BOWL URINAL RANGES (in vitreous china)

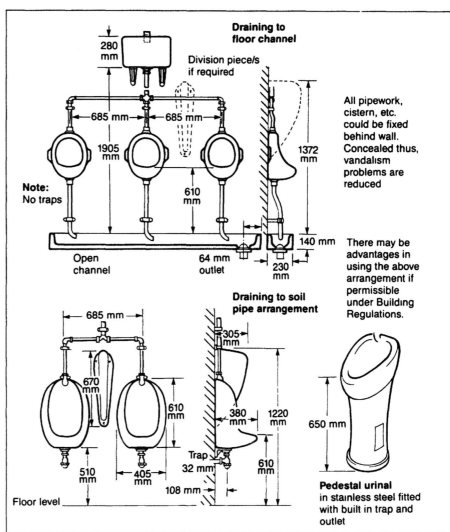

280 mm

Draining to floor channel

Division piece/s if required

685 mm — 685 mm

1905 mm

610 mm

Note: No traps

Open channel

64 mm outlet

1372 mm

140 mm
230 mm

All pipework, cistern, etc. could be fixed behind wall. Concealed thus, vandalism problems are reduced

There may be advantages in using the above arrangement if permissible under Building Regulations.

685 mm

670 mm

610 mm

510 mm

405 mm

Draining to soil pipe arrangement

305 mm

380 mm

1220 mm

Trap 32 mm

108 mm

610 mm

Floor level

650 mm

Pedestal urinal in stainless steel fitted with built in trap and outlet

OUTLETS AND TRAPS

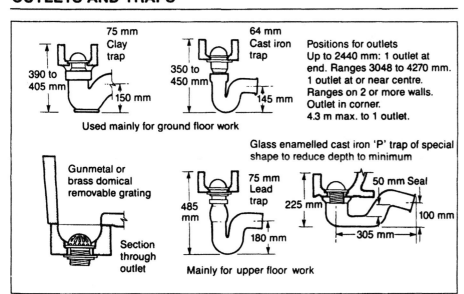

75 mm Clay trap

390 to 405 mm

150 mm

Used mainly for ground floor work

Gunmetal or brass domical removable grating

Section through outlet

64 mm Cast iron trap

350 to 450 mm

145 mm

Positions for outlets
Up to 2440 mm: 1 outlet at end. Ranges 3048 to 4270 mm. 1 outlet at or near centre. Ranges on 2 or more walls. Outlet in corner. 4.3 m max. to 1 outlet.

Glass enamelled cast iron 'P' trap of special shape to reduce depth to minimum

75 mm Lead trap

485 mm

180 mm

Mainly for upper floor work

225 mm

50 mm Seal

100 mm

305 mm

ANTI SPLASH

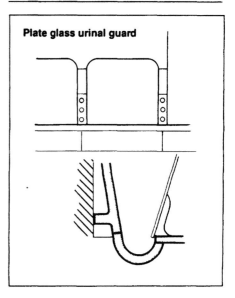

Plate glass urinal guard

4

VALVELESS W.C. CISTERNS

Operated by means of a syphon, this type is termed a 'waste water preventer', and is generally the only type permitted in the U.K. It has a capacity of 9 litres.

Dual flush is available to provide a full flush or partial flush, which enables a 50 per cent saving of water to be made. It is normally operated by pulling the lever handle sharply, and allowing it to return. Full flush is obtained by holding down the handle normally.

Ball valves are available to suit high or low-pressure supply — 3 bar and above would need a high pressure seat; below 3 bar, a low-pressure seat should be fitted. Fittings are available to control noise on bottom inlet cisterns; the fitting of a control valve on cistern with normal side inlet will reduce flow, reduce noise and help with maintenance.

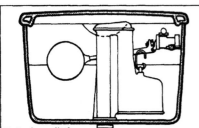

Valveless fittings
with plastic siphon, ball valve, float and bottom or side inlet.
Can be supplied with low height siphon to act as internal overflow where this is permitted by the Water Authority.

Complies with U.K. Water Authority Regulations and Model Byelaws.

AUTOMATIC FLUSHING

The Cistermiser control valve, made by the company of the same name, was developed to restrict the flow of water to automatic flushing cisterns at those times when the facilities were not in use.

The automatic flushing cistern was known to waste large quantities of water every year because they continue to flush even at times when the sanitary appliances were not in use — at night or weekends.

This was recognised by water authorities and was subsequently responsible for the introduction of byelaws for the control of automatic flushing cisterns.

Method of operation
The valve is substituted for the trickle cock mechanism and is activated by a reduction in the water pressure caused by use of an appliance. This will then allow a pre-set quantity of water to pass through to a given number of appliances served.

At times of non-use of the buildings such as weekends and holidays, the even water pressure will prevent water passing through the valve into the cistern. The sensitivity of the valve is such that it will be activated at about 5 per cent drop of the supply pressure. Once installed and adjusted, it requires no further maintenance.

It is made from vandal resistant plastic, apart from an adjusting screw and brass bellows.

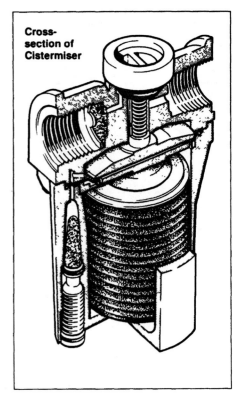

Cross-section of Cistermiser

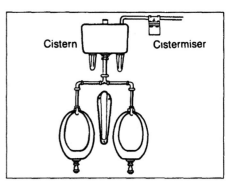

Cistern   Cistermiser

VALVE TYPE W.C. CISTERNS

These are available as an alternative to 'waste water preventers', where water authorities' regulations permit.

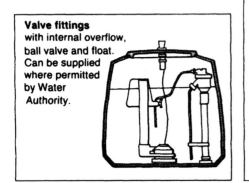

Valve fittings
with internal overflow, ball valve and float. Can be supplied where permitted by Water Authority.

Automatic flushing system

Automatic flushing cisterns are available, and the diagram features one from the Armitage Shanks catalogue. This unit works on an action initiated by siphonage. The capacity of the flush should not exceed 4.5 litres (or 13.5 litres per hour) for each wall urinal, stall or 700 mm or slab. Each automatic cistern must be fitted with a time switch or similar device.

The provision of volume regulating valves is becoming increasingly popular due to the ability, if fitted, to provide temporary isolation of appliance for maintenance or repair, avoiding the shutting off of the supply to other appliances.

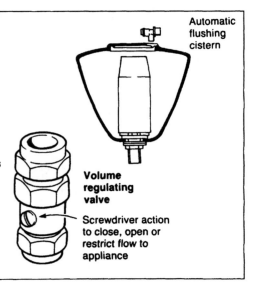

Automatic flushing cistern

Volume regulating valve

Screwdriver action to close, open or restrict flow to appliance

Sanitation: Appliances

TROUGH TYPE W.C. CISTERNS

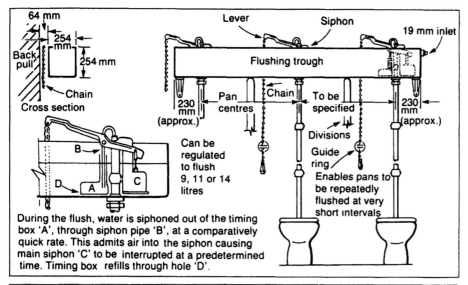

Back pull
64 mm
254 mm
254 mm
Chain
Cross section

Lever
Siphon
19 mm inlet
Flushing trough
230 mm (approx.)
Pan centres
Chain
To be specified
Divisions
Guide ring
Enables pans to be repeatedly flushed at very short intervals
230 mm (approx.)

Can be regulated to flush 9, 11 or 14 litres

During the flush, water is siphoned out of the timing box 'A', through siphon pipe 'B', at a comparatively quick rate. This admits air into the siphon causing main siphon 'C' to be interrupted at a predetermined time. Timing box refills through hole 'D'.

FLUSH PIPES

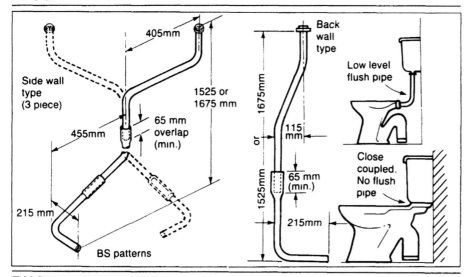

Side wall type (3 piece)
405mm
455mm
65 mm overlap (min.)
215 mm
BS patterns

1525 or 1675 mm

Back wall type
Low level flush pipe
1675mm
1525mm
115 mm
65 mm (min.)
215mm

Close coupled. No flush pipe

DUCTED CISTERNS

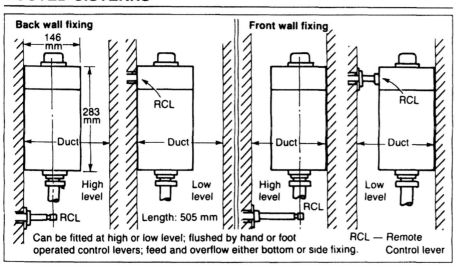

Back wall fixing
146 mm
283 mm
Duct
High level
RCL

Duct
Low level
Length: 505 mm

Front wall fixing
Duct
High level
RCL

Duct
Low level
RCL

RCL — Remote Control lever

Can be fitted at high or low level; flushed by hand or foot operated control levers; feed and overflow either bottom or side fixing.

FLUSHING VALVES

These are suitable for high usage W.C. installations allowing repeated flushing without any delay. It is fitted in place of a conventional flushbend cistern and is available in both exposed and concealed patterns. The valve has a self-closing action after release of a pre-determined volume of water.

Water authorities in the UK will not generally accept flush valves, except under social special conditions — and permission should be sought prior to installation. Supplies should be tank-fed to give the required flow rate through the valve of 114 litres per minute.

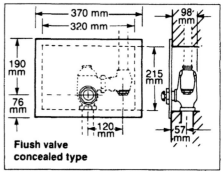

370 mm
320 mm
98 mm
190 mm
215 mm
76 mm
120 mm
57 mm

Flush valve concealed type

CONTROLS

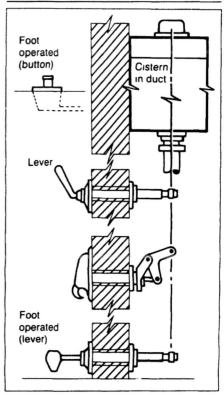

Foot operated (button)
Cistern in duct
Lever
Foot operated (lever)

Miscellaneous Appliances

SLOP HOPPERS, SINK AND BED PAN WASHERS

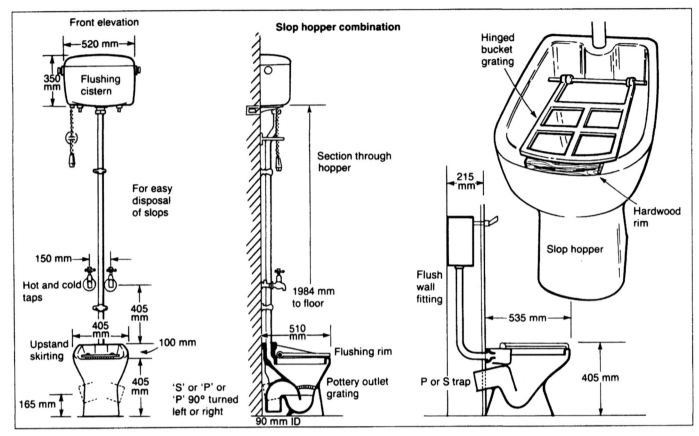

Front elevation

520 mm

Flushing cistern

350 mm

For easy disposal of slops

150 mm

Hot and cold taps

405 mm

405 mm

100 mm

Upstand skirting

405 mm

165 mm

'S' or 'P' or 'P' 90° turned left or right

Slop hopper combination

Section through hopper

1984 mm to floor

510 mm

Flushing rim

Pottery outlet grating

90 mm ID

Hinged bucket grating

215 mm

Hardwood rim

Slop hopper

Flush wall fitting

535 mm

P or S trap

405 mm

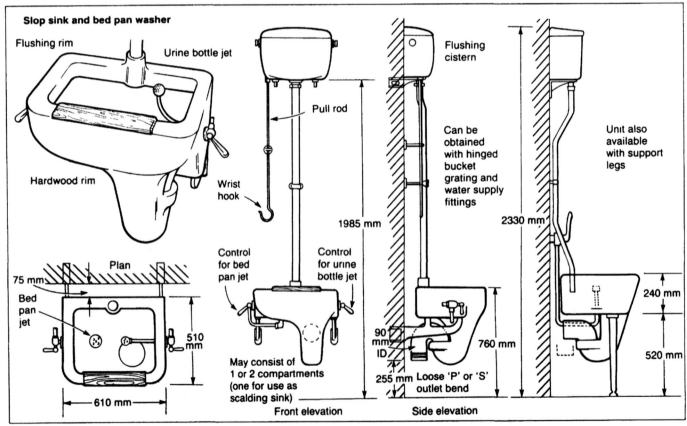

Slop sink and bed pan washer

Flushing rim

Urine bottle jet

Hardwood rim

Pull rod

Wrist hook

Plan

75 mm

Bed pan jet

510 mm

610 mm

Control for bed pan jet

Control for urine bottle jet

May consist of 1 or 2 compartments (one for use as scalding sink)

Front elevation

Flushing cistern

Can be obtained with hinged bucket grating and water supply fittings

1985 mm

90 mm ID

255 mm

Loose 'P' or 'S' outlet bend

760 mm

Side elevation

Unit also available with support legs

2330 mm

240 mm

520 mm

Sanitation: Appliances

ABLUTION APPLIANCES

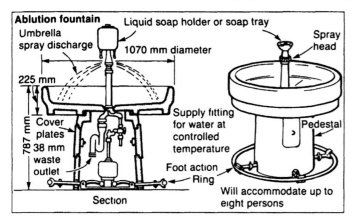

Ablution fountain Liquid soap holder or soap tray
Umbrella spray discharge
1070 mm diameter
Spray head
225 mm
787 mm
Cover plates
38 mm waste outlet
Section
Supply fitting for water at controlled temperature
Foot action Ring
Pedestal
Will accommodate up to eight persons

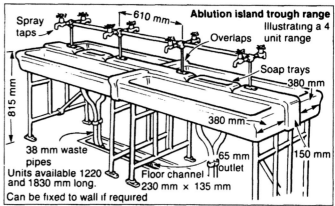

Ablution island trough range
Illustrating a 4 unit range
Spray taps
610 mm
Overlaps
Soap trays
380 mm
815 mm
380 mm
38 mm waste pipes
Units available 1220 and 1830 mm long.
Can be fixed to wall if required
Floor channel 230 mm × 135 mm
65 mm outlet
150 mm

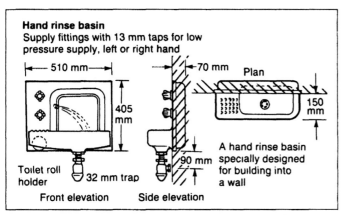

Hand rinse basin
Supply fittings with 13 mm taps for low pressure supply, left or right hand
510 mm
405 mm
70 mm
Plan
150 mm
Toilet roll holder
32 mm trap
90 mm
Front elevation
Side elevation
A hand rinse basin specially designed for building into a wall

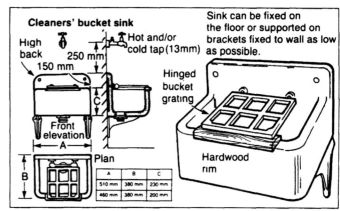

Cleaners' bucket sink
Sink can be fixed on the floor or supported on brackets fixed to wall as low as possible.
High back
250 mm
Hot and/or cold tap (13mm)
150 mm
C
Front elevation
A
Plan
B
Hinged bucket grating
Hardwood rim

A	B	C
510 mm	380 mm	230 mm
460 mm	380 mm	200 mm

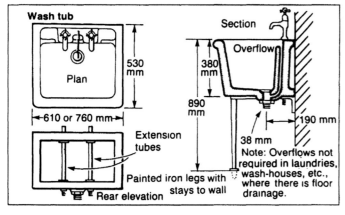

Wash tub
Plan
530 mm
Section
Overflow
380 mm
890 mm
610 or 760 mm
190 mm
Extension tubes
38 mm
Painted iron legs with stays to wall
Rear elevation
Note: Overflows not required in laundries, wash-houses, etc., where there is floor drainage.

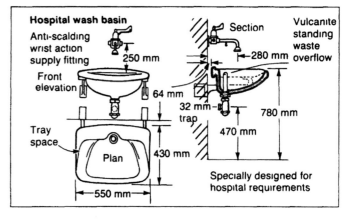

Hospital wash basin
Anti-scalding wrist action supply fitting
250 mm
Section
Vulcanite standing waste overflow
Front elevation
64 mm
280 mm
32 mm trap
780 mm
Tray space
Plan
470 mm
430 mm
550 mm
Specially designed for hospital requirements

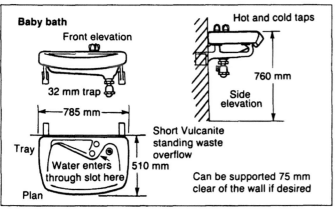

Baby bath
Hot and cold taps
Front elevation
32 mm trap
760 mm
785 mm
Side elevation
Tray
Water enters through slot here
510 mm
Plan
Short Vulcanite standing waste overflow
Can be supported 75 mm clear of the wall if desired

Definitions Authoritative Meanings Extracted from Various

EXPRESSION	DEFINITION	AUTHORITY
Chemical closet	A closet having a receptacle for the reception of faecal matter and its deodorisation by the use of suitable chemicals.	Sanitary Conv. Regs. 1964 S.I.966 Shops and Offices
Earth closet	A closet having a movable receptacle for the reception of faecal matter and its deodorisation by the use of earth, ashes or chemicals, or by other methods.	Public Health Act 1936 and 1961
Prejudicial to health	Injurious, or likely to cause injury, to health.	Public Health Act 1936 Section 343 page 210
Privy	Included in the interpretation of closet.	Public Health Act 1936 Section 90 page 66
Sanitary accommodation	A room or space constructed for use in connection with a building and which contains water closet fittings or urinal fittings, whether or not it also contains other sanitary or lavatory fittings.	Building Regulations 1972 Part P3 page 108
Sanitary appliance	An appliance fitted to a drainage system for the collection and discharge of foul or waste matter.	
Sanitary conveniences	Includes urinals, water closets, earth closets, privies, ashpits, and any similar convenience.	Factories Act 1961 Section 176 page 115
Sanitary conveniences	Means closets and urinals	Public Health Act 1936 Section 90 page 66
Sanitary convenience	A water closet, urinal, chemical closet or similar convenience.	Food Hygiene Regulations S.I.1172
Soil appliance	A sanitary appliance for the collection and discharge of excretory matter	C.P.305:52 Sanitary Appliances page 8
Trough	A trough measuring internally at least four feet over its longest or widest part with a smooth impervious surface and fitted with an unplugged waste pipe and having a supply of warm water laid on at points above the trough and at suitable intervals of not more than 0.6m.	Washing Facilities Regulations 1964 S.I.965 Shops and Offices page 1
Unit of trough or washing fountain accommodation	Two feet of length of a trough or, in the cases of circular or oval troughs and washing fountains, two feet of the circumference of the trough or fountain.	As above
Urinal	A urinal which is connected to a drainage system and which has provision for flushing from a supply of clean water either by the operation of mechanism or by automatic action.	Sanitary Conv. Regulations 1964 S.I.966 Shops and Offices page 1
Wash basin	A fixed basin with a smooth impervious surface, having a supply of clean running hot and cold or warm water and fitted with a waste pipe and (except where the supply of water is from a spray tap) with a plug.	Washing Facilities Regulations 1964 S.I.965 Shops and Offices page 1
Wash bowl	Includes any water container suitable for use as a washing facility.	As above
Washing fountain	A washing fountain measuring internally at least 0.9 m over its widest part, with a smooth impervious surface and fitted with an unplugged waste pipe and having a supply of running warm water.	As above
Waste appliance	A sanitary appliance for the collection and discharge of water after use for ablutionary, culinary and other domestic purposes.	C.P.305:52 Sanitary Appliances page 8
Waste appliance	Includes a slipper bath, lavatory basin*, bidet, domestic sink, cleaner's bucket sink, drinking fountain, shower tray, wash fountain, washing trough and wash tub (* term 'washbasin' is preferred — see BS.4118).	Building Regulations 1972 Part N2 page 101

Sanitation: Appliances

EXPRESSION	DEFINITION	AUTHORITY
Water closet (W.C.)	Washdown type. A W.C. in which the contents of the pan are removed by a flush of water discharged into the pan.	
	Pedestal type, siphonic. A. W.C. in which the contents of the pan are removed by siphonage.	BS.6465
	Wall hung type. A washdown pan supported from the wall by a projecting corbel at flushing line level, or by brackets, thus providing freedom of floor space under the pan for cleansing purposes.	As above
Water closet	A closet which has a separate fixed receptacle connected to a drainage system and separate provision for flushing from a supply of clean water either by the operation of mechanism or by automatic action.	Public Health Act 1936 Section 90 page 66 and Sanitary Conv. Regs. 1964 S.I.966 Shops and Offices page 1
Water fittings	Includes pipes (other than mains), taps, cocks, valves, ferrules, meters, cisterns, baths, water closets, soil pans and other similar apparatus used in connection with the supply and use of water.	Water Act 1945 page 55 and 72 and Model Water Byelaws (1966 Edition) page 4

Important note: the above definitions are the meanings assigned to certain expressions for the purpose of the document referred to as the 'authority'. Great care must be taken when using them out of the context for which they were intended but nevertheless, they serve as a useful guide for general interpretation.

CONSTRUCTION REQUIREMENTS

Brackets for building into a wall should include a lugged portion which may be slotted to provide a key for mortar. Brackets for screwing to a wall should be provided with a back fixing plate. Supports consisting of legs with horizontal straps should have a wall fixing flange on the latter. The leg should be rigidly and securely fixed to the strap. Strap and leg supports without fixing studs should be tied together by a cross member, either adjacent to the wall or at the outer extremity. Leg supports may be solid drawn tubes or castings and may terminate in a screw flange or alternative types of dowel flange. If towel rails are provided with strap and leg supports, the ends adjacent to the wall should be secured to the wall fixing flange and not independently attached to the wall.

FIXING HEIGHTS

Wash basins:
Traditional recommended height for use by adults is 785 mm or 800 mm to the rim. A basin fixed at a height of 700 mm would be appropriate where children are catered for.

W.C.s:
Traditional height for adult use is 400 mm. Lower heights are available for children, as well as higher units for the elderly or disabled — 450 mm or more.

Urinal pods:
The preferred rim height for urinal bowls is 600 mm for adult use; where younger boys are present, one pod at least should be at a lower height of about 500 mm from floor level to the rim.

STANDARD BRACKETS AND SUPPORTS

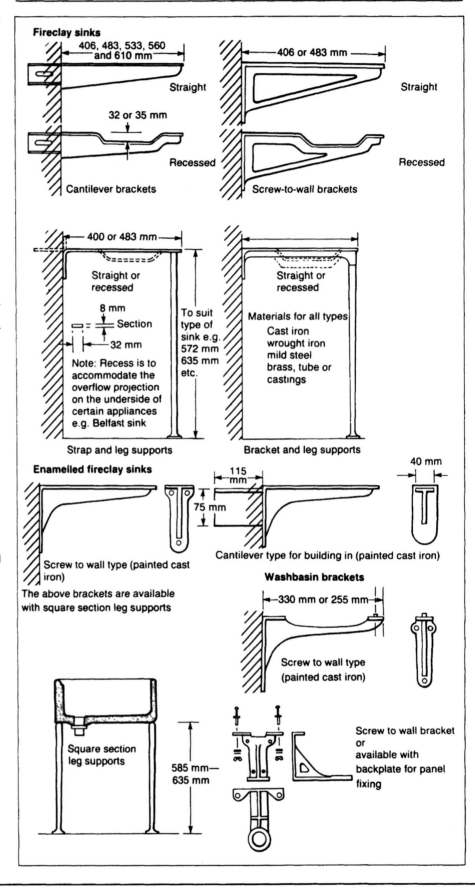

Fireclay sinks

406, 483, 533, 560 and 610 mm — Straight
32 or 35 mm — Recessed
Cantilever brackets

406 or 483 mm — Straight
Recessed
Screw-to-wall brackets

400 or 483 mm — Straight or recessed
8 mm — Section — 32 mm
Note: Recess is to accommodate the overflow projection on the underside of certain appliances e.g. Belfast sink
Strap and leg supports

Straight or recessed
To suit type of sink e.g. 572 mm 635 mm etc.
Materials for all types
Cast iron wrought iron mild steel brass, tube or castings
Bracket and leg supports

Enamelled fireclay sinks
115 mm
75 mm
40 mm
Screw to wall type (painted cast iron)
Cantilever type for building in (painted cast iron)

The above brackets are available with square section leg supports

Washbasin brackets
330 mm or 255 mm
Screw to wall type (painted cast iron)

Square section leg supports
585 mm—635 mm
Screw to wall bracket or available with backplate for panel fixing

STANDARD BRACKETS AND SUPPORTS

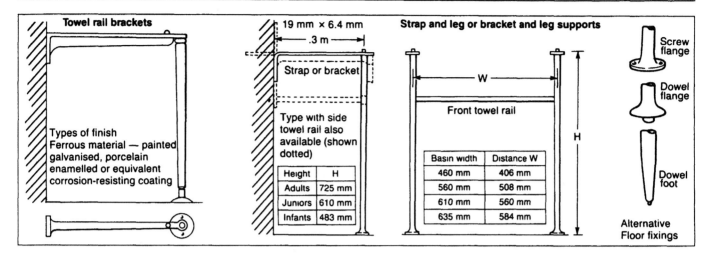

Towel rail brackets

Types of finish
Ferrous material — painted galvanised, porcelain enamelled or equivalent corrosion-resisting coating

19 mm × 6.4 mm
.3 m

Strap or bracket

Type with side towel rail also available (shown dotted)

Height	H
Adults	725 mm
Juniors	610 mm
Infants	483 mm

Strap and leg or bracket and leg supports

W

Front towel rail

H

Basin width	Distance W
460 mm	406 mm
560 mm	508 mm
610 mm	560 mm
635 mm	584 mm

Screw flange

Dowel flange

Dowel foot

Alternative Floor fixings

MISCELLANEOUS BRACKETS AND SUPPORTS

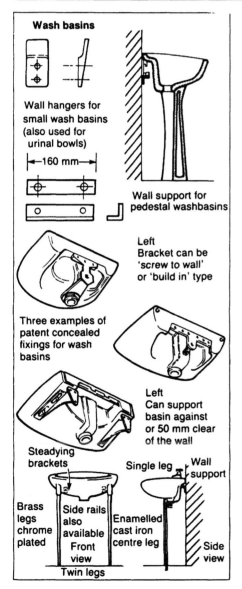

Wash basins

Wall hangers for small wash basins (also used for urinal bowls)

|←160 mm→|

Wall support for pedestal washbasins

Left
Bracket can be 'screw to wall' or 'build in' type

Three examples of patent concealed fixings for wash basins

Left
Can support basin against or 50 mm clear of the wall

Steadying brackets

Single leg

Wall support

Brass legs chrome plated

Side rails also available

Enamelled cast iron centre leg

Front view

Side view

Twin legs

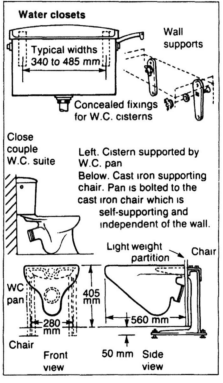

Water closets

Typical widths 340 to 485 mm

Wall supports

Concealed fixings for W.C. cisterns

Close couple W.C. suite

Left. Cistern supported by W.C. pan
Below. Cast iron supporting chair. Pan is bolted to the cast iron chair which is self-supporting and independent of the wall.

Light weight partition

Chair

WC pan

405 mm

Chair

←280 mm→

560 mm

Front view

50 mm

Side view

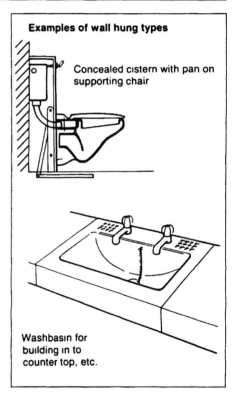

Examples of wall hung types

Concealed cistern with pan on supporting chair

Washbasin for building in to counter top, etc.

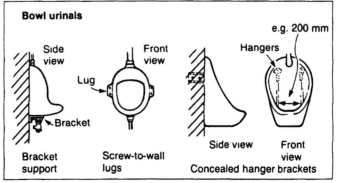

Bowl urinals

Side view

Front view

Lug

Bracket

e.g. 200 mm

Hangers

Side view

Front view

Bracket support

Screw-to-wall lugs

Concealed hanger brackets

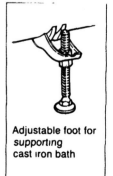

Adjustable foot for supporting cast iron bath

8

Tabulated Information Sanitary Appliances

APPLIANCE	B.S. No.	PURPOSE	MATERIALS
Water closet flushing cisterns	1125:1973	Flushing W.C. pans, slop hoppers, slop sinks, etc.	Cast iron, ceramic ware, pressed steel, lead or copper lined wood, composition and other materials (including plastics)
Flush pipes	1125:1973	Connecting cisterns to water closet pans, etc.	Enamelled or galvanised steel, lead copper, copper alloy, uPVC, or suitable copolymer
Automatic Flushing Cisterns for urinals	1876:1972 (1977)	Flushing urinals	As for BS.1125
Ceramic washdown W.C. pans	5503:1981	Removal of human wastes (solid and liquid)	Caneware 1 and 2, earthenware (ordinary or heavy), fireclay, stoneware, vitreous china (ordinary or heavy duty)
W.C. seats (plastic)	1254:1981	Hygienic	Type 1 specified phenolic or amino-plastic. Type 2 any material satisfying tests in BS.
Portable closets for use with chemicals	2081: Part 1: 1980	Removal of human wastes (solid and liquid)	Container to be of non-absorbent material. Any outer case can be ditto.
Ceramic wash basins and pedestals	1188:1974	Personal ablution	Fireclay or vitreous china (ordinary or heavy grade). (Also available made of plastic material)
Metal lavatory basins	1329:1974	Personal ablution	Cast iron or pressed steel (vitrified porcelain enamelled). Stainless steel
Cast iron baths for domestic purposes	1189:1972 (1980)	Personal ablution	Cast iron porcelain enamelled (also made of fireclay but non BS). Panels — superhard board, steel, asbestos, cement, plastic etc.
Sheet steel baths for domestic purposes	1390:1972	Personal ablution	Pressed steel vitreous enamelled
Fireclay sinks	1206:1974	Culinary, laundry and other domestic purposes	Fireclay, white inside and out, glaze fused by firing
Metal sinks for domestic purposes	1244: Part 2: 1982	Culinary, laundry, etc. (for fixing to sink units)	BS materials as for BS.1329 (see above) (Available made of fireclay, 'Perspex', polypropylene, etc.)
Baths made from cast acrylic sheet	4305:1972 (1977)	Personal ablution	
Slab urinals (stainless steel)	4880: Part 1: 1973	Removal of human liquid waste	Stainless steel
Bidets	5505: Parts 1, 2 and 3, 1977	Cleansing lower organs of the body (excretory etc.)	Choice of certain ceramic materials as for water closet pans
Shower trays	6340: Parts 5, 6, 7 and 8: 1983	Personal ablution	Choice of certain ceramic materials, enamelled pressed steel, Perspex, etc.
Vitrified clay bowl urinals (rimless type)	5520:1977	Disposal of human liquid waste	Vitreous china

Sanitation: Appliances

MAIN DIMENSIONS

Capacity — 4 litre, 5 litre and 9 litre ±5% single flush. (11 and 14 litre also available — non BS.). Approximate width (91) 510 mm-560 mm. Height to top of cistern low level maximum 1070 mm: close couple e.g. 740-790 mm: high level to suit BS.1525 or 1675 mm long flush pipes

For pans to BS.1213 — high level — minimum internal diameter 32 mm and effective height 1525-1675 mm. Low level minimum internal diameter 35 mm.

Nominal capacity 4.5l per stall served or equivalent length of slab. Nominal sizes: 4.5l*, 9l*, 13.5l*, 18l+, 22.5l‡ and 27l‡.

Normal height 406 mm (pans are also made 255, 305 and 355 mm for children of various ages). Width (approximately) 355 mm. Overall length: 520 mm-635 mm. Outlet heights: floor to centre line of 'P' trap 190 mm. Floor to end of 'S' trap 19 mm.

Minimum width at widest point 380 mm. Centre line of hinges to front edge — minimum 445 mm; maximum 476 mm. Finished thickness — 20 mm (minimum)

Total height to top face of seat (less cover) minimum 356 mm, maximum 457 mm. Maximum capacity 32 litre.

BS overall plan sizes: 635 × 460 mm or 560 × 406 mm (for confined spaces). When fixed to pedestal, top of rim at front of basin to be 785 mm above floor level.

Plan sizes as BS.1188

Length: co-ordinating size 1700 mm (other co-ord. sizes = n × 100 mm. Work size 1694 mm. Width: co-ordinating size 700 mm (other co-ord. sizes = n × 100 mm. Work size 697 mm. Height: co-ordinating size = n × 50 mm (n = any natural number including unity)

As for BS.1189

A. Reversible sinks without shelves (Belfast)
B. Sinks with back shelves.
C. Sinks with 70 mm ledge at back.‡
D. Combination sinks with integral drainer.
F. As above, with flange for fixing in cabinets.
G. Double (two compartment) sinks.

Type A. Single bowl, single drainer, RH or LH.
Type B. Single bowl, double drainer.
Type C. Single bowl, double drainer, RH or LH.

	Type A		Type B		Type C		
Co-ord.	1000	1200	1500	1800	1500	1800	L
Size (mm)	600	600	600	600	600	600	W

Dimensions as for cast iron and pressed steel. Two types of baths are included:- 1. 8 mm cast acrylic sheet.
2. Reinforced with resin and glass fibre

Height above tread — 1050 mm. Length: n = 600 mm. Height of cistern above tread 1800 mm (minimum) 2000 mm (maximum). Single or integrated multi units

Height from floor level — about 380 mm. Approximate overall plan dimensions — width 330 mm, front to rear 530 mm to 640 mm

Examples of outside dimensions 610 mm × 610 mm × 180 mm, 760 mm × 760 mm × 180 mm. Obtainable with or without overflows. Can drain to floor channel if required

Height is maximum 600 mm; depth is 380 mm maximum (overall projection of bowl from back surface); width of bowl at widest point is 400 mm maximum

OUTLET DETAILS

Flush pipe connections — high level 32 mm (min.) low level 38 mm (min.) 19 mm overflow if fitted with ½" inlet

Bore used for 1 m test length = *19 mm, + 25 mm, ‡32 mm.

Trap is integral, internal diameter 89 mm. Seal depth 50 mm. Trap 'S' or 'P' straight or turned left or right.

Should have no outlet or overflow

Requires provision of separate trap. Building Regulations minimum waste size 32 mm. Integral overflow

As above and integral overflow provided unless otherwise stated

Requires provision of separate trap. Waste size recommended for domestic use — 38 mm. No overflow hole unless ordered

As above

As above but overflow are integral except Type C‡ is without overflow

As above and overflows are provided unless otherwise stated. Choice of two outlet positions

Requires provision of separate trap. Recommended waste size 38 mm

Requires provision of separate trap. Waste hole 80 mm ±2 mm diameter

Normally 32 mm. Requires provision of separate trap ('S' or 'P')

Normally 38 mm from centre or corner or side of tray. Separate trap required

Diameter of waste outlet is 50 mm

NOTES

Flush rate maximum
H.L. — 91 in 5 secs
L.L. — 91 in 6 secs
Dual flush, choice of 4.5 or 91.
Short — 'let go'; Long — 'hold'.

Single or double telescopic

Usually set to operate average 20 minute intervals

Approximate projection from rear wall when fixed:-
H.L. 686 mm
L.L. 711 mm-760 mm
C.C. 760 mm
Concealed cistern 520 mm

Shall be flat on the underside — not recessed

Vent pipe may be attached but shall not be essential. Test load: 127 kg

Use cover pieces or allow space 75 mm between when fixed in ranges. Minimum weights specified

Minimum thicknesses of materials specified

Should be as flat bottomed as possible and self-draining when rim is level. Single handgrip fitted to back roll unless not specified

Minimum thickness 1.6 mm. Other notes — as above

Drainer height to be 915 mm from floor and height of sink to be adjusted to suit. Hardwood pads to order.

Height to top of front edge to be 915 mm. Bowl size example.
Length 430 mm
Width 360 mm
Depth 180 mm
Also round bowls

Shall have no overflow unless specified. Handgrips may be provided

With or without divisions. Channel fall to outlet: 1/100

Appliance has a secondary use — feet washing. Normally made with integral overflow

Often referred to as "footbaths" but requires pipework arranged for this purpose with plug and overflow

Stainless steel and plastic types are also available, not covered by BS

Key H.L. — High level   L.L. — Low level
     C.C. — Close couple   L   — Length   W — Width
     L.H. — Left hand   R.H.— Right hand

9

APPLIANCE	PURPOSE	MATERIALS
Sinks domestic type London pattern	Culinary, laundry and general purpose	Choice of certain ceramic materials
Drinking fountains	Drinking from directly	Choice of ceramics, enamelled cast iron, brass, chrome, plated etc.
Drip sinks	Enables low level filling and emptying of buckets	Choice of certain ceramic materials
Flushing valves	Mainly for flushing water closet pans	Brass, chromium plated
Siphonic water closet pans	Removal of human wastes (solid and liquid)	Choice of certain ceramic materials
Slop Hoppers (or housemaids' closet)	Removal of 'slops' which may consist of dirty or fouled water, human (solid and liquid) waste	Choice of certain ceramic materials
Slop sinks		Choice of certain ceramic materials
Squat type (Eastern or Asian) closets	Removal of human wastes (solid and liquid)	Choice of certain ceramic materials
Trough type flushing cisterns	Flushing W.C. pans (when short interval flushing required e.g. schools, etc.)	Galvanised mild steel or certain plastic materials
Trough (ablution)	Personal ablution	Choice of certain ceramic materials or porcelain, enamelled cast iron
Wash fountains (circular)	Personal ablution	Ceramic material or stainless steel

Sanitation: Appliances

MAIN DIMENSIONS

Examples of sizes that may be available (some sizes to be withdrawn).
760 × 460 × 255 mm deep.
760 × 460 × 225 mm deep.
610 × 460 × 225 mm deep.
610 × 460 × 150 mm deep.
460 × 380 × 205 mm deep.
430 × 330 × 150 mm deep.
(Note: All outside dimensions)

Outside dimensions of 2 examples — width 380mm, projection 280 mm, height 180 mm. width 230 mm, projection 360 mm, height 150 mm

Outside dimensions of 2 examples — 510 × 380 × 230 mm deep. 460 × 380 × 200 mm deep. Can be fixed at floor level or raised clear of the floor by several cm.

Wall space occupied — 150mm² approximately. This fitting is useful for 'short interval' flushing. Working pressure and supply pipe size require careful consideration (likely to be about 32 mm)

Height — 380 to 410 mm. Width — 380 mm. Outlet height varies with design. Projection from rear wall when fixed — double trap 710-810 mm. Single trap — 680 mm. If flush pipe swannecked to wall — 660 mm

Height to front edge — approximately 410 mm (certain models can be set in the floor). Typical plan dimensions 410 mm wide and 510 mm projection from rear wall. Corbel types are available.

Fixing height to front edge — 760 mm. Typical plan dimensions 610 mm wide and 510 mm projection from rear wall. Usually incorporates urine bottle and bed pan washer

Plan measurements overall — width 430 mm to 760 mm and front to rear 610-760 mm. Projection from rear wall (fixed) 760-860 mm

Unit suitable for 3 W.C.s at 810-910 mm centres supported on brackets or compartment partitions, e.g. 240 mm high, 225 mm wide, 2.1 m long. Some sections 'bolt' together as continuous trough

Length (for two persons) 1220 mm. Length (for three persons) 1070 mm special design width 330-380 mm, depths 150-180 mm. Fixing heights as for wash basins (approximately). Can be fixed end to end and back to back

Diameter on plan — 1070 mm (will accommodate 8 persons). Fixing height when on pedestal — 810 mm

OUTLET DETAILS

Normally 38 mm. Requires the provision of separate trap — 'S', 'P' etc.

Normally 32 mm. Requires provision of separate trap

Normally 32 mm. Requires the provision of separate trap

32 mm flush pipe. Length of flush pipe — approximately 0.3 m

Trap is integral. Single or double trap Internal diameter 76 to 83 mm but varies with designs

89 mm integral trap. 'S' or 'P', pointing to rear or turned left or right

89 mm integral or part integral trap. 'S' or 'P' normal or turned any angle

Trap integral 'S' or 'P'. Internal diameter — 89 mm

Flush pipes normally 32 mm

Normally 38 mm. Separate trap required or may discharge to floor channel

As for troughs (ablution)

NOTES

London pattern sinks have no integral overflow. Fix as for sinks made to BS.1206

For wall or pedestal mounting. User should drink direct from jet

Should have hinged grating on top and hardwood pad set in front edge

Only where permissible may a flush valve be used to replace a cistern. Supply to be from strictly separate storage cistern

Examples of heights to top of cistern — 790 mm — 1 m. Examples of cistern widths
9 litre — 510 mm
11 litre — 530 mm
14 litre — 560 mm

Flushed by cistern. Should have hinged grating on top and loose grating above water level and hardwood pad

Flushed by cistern. Has hardwood pad inset and may have hinged grating

Depth required from floor level varies with design e.g. 360-520 mm

Will flush 9, 11 or 14 litre per operation at very short intervals according to the setting of the timing arrangement

Can cope with peak demand periods better than individual basins, therefore they are ideal for factories, schools etc.

As for troughs. Requires minimum 3m head to operate spray head satisfactorily (approximately .023 l/sec)

Key   H.L. — High level   L.L. — Low level
      C.C. — Close couple   L   — Length   W — Width
      L.H. — Left hand   R.H.— Right hand

10

Design Allowances and Space

SPACE REQUIREMENTS FOR COMMON APPLIANCES

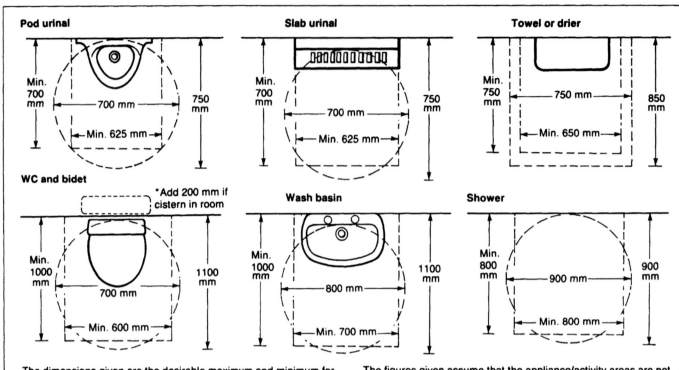

The dimensions given are the desirable maximum and minimum for given appliances and not absolute.

The figures given assume that the appliance/activity areas are not bounded at the sides by walls, but are adjoining other similar areas, where this occurs a distance of 400 mm is recommended from the centre of the appliance to the wall.

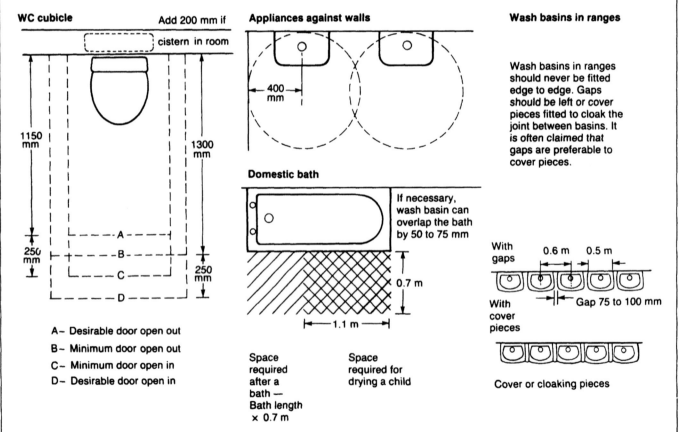

WC cubicle

A– Desirable door open out
B– Minimum door open out
C– Minimum door open in
D– Desirable door open in

Appliances against walls

Domestic bath

If necessary, wash basin can overlap the bath by 50 to 75 mm

Space required after a bath — Bath length × 0.7 m

Space required for drying a child

Wash basins in ranges

Wash basins in ranges should never be fitted edge to edge. Gaps should be left or cover pieces fitted to cloak the joint between basins. It is often claimed that gaps are preferable to cover pieces.

With gaps

With cover pieces

Gap 75 to 100 mm

Cover or cloaking pieces

Sanitation: Appliances

EXAMPLES OF BATHROOM LAYOUTS

Bathrooms A, B and C are extracted from MoHLG Design Bulletin No. 8 'Dimensions and Components for Housing'. The dimensions recommended are based on the clearances and user requirements given in various authoritative publications. (Note: Entrances not shown)

Shaded portions indicate duct space. Similar arrangements are possible without duct space if such a design is required

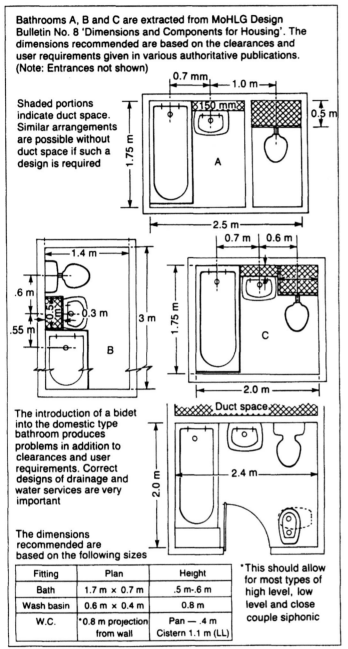

The introduction of a bidet into the domestic type bathroom produces problems in addition to clearances and user requirements. Correct designs of drainage and water services are very important

The dimensions recommended are based on the following sizes

Fitting	Plan	Height
Bath	1.7 m × 0.7 m	.5 m-.6 m
Wash basin	0.6 m × 0.4 m	0.8 m
W.C.	*0.8 m projection from wall	Pan — .4 m Cistern 1.1 m (LL)

*This should allow for most types of high level, low level and close couple siphonic

Miscellaneous design data
Estimating total space required per fitment in factories (includes allowances for fitment, user and circulation)
Per water closet — 1.8m². Per wash basin — 0.75 m². Per urinal position — 1.2 m².
Estimating total space required per employed person. Washing only — 0.15 m². For all purposes — 0.33 m² (minimum) but a better standard could be 0.5 m² to 0.6 m² particularly if cloakroom facilities are to be provided and the space is available.
When providing a 'Powder Room' for ladies, mirrors should not be sited over the wash basins but elsewhere in the room with a shelf beneath. Passing space is required and user space for standing or sitting.
Note: Special compartment sizes are required for handicapped persons (see BS.5810:1979).

MISCELLANEOUS LAYOUTS

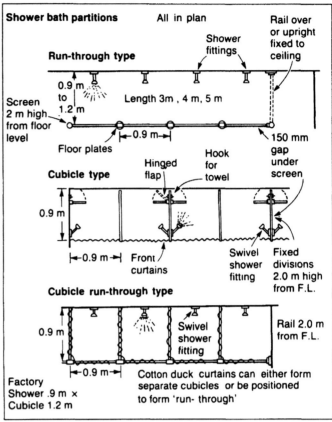

Cotton duck curtains can either form separate cubicles or be positioned to form 'run-through'

Factory Shower .9 m × Cubicle 1.2 m

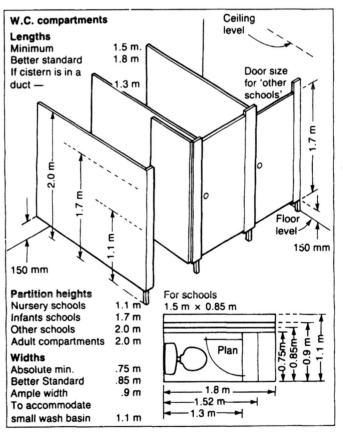

W.C. compartments
Lengths
Minimum	1.5 m.
Better standard	1.8 m
If cistern is in a duct —	1.3 m

Door size for 'other schools'

Partition heights
Nursery schools	1.1 m
Infants schools	1.7 m
Other schools	2.0 m
Adult compartments	2.0 m

Widths
Absolute min.	.75 m
Better Standard	.85 m
Ample width	.9 m
To accommodate small wash basin	1.1 m

For schools
1.5 m × 0.85 m

Materials A Selection of Materials Used to Manufacture

POINTS CONSIDERED	CERAMIC MATERIALS	CAST IRON
BASIC MATERIAL	**Fireclay.** Semi-porous yellow (buff) refactory clay of great strength. Can be used for large appliances for which vitreous china could not be used. **Vitreous China.** White non-porous clay body of very fine texture. Strong, best quality ware which should not craze, stain or deteriorate in any way.	**Mainly concerning baths.** The cast iron used in the manufacture of baths shall be remelted soft tough grey iron of a suitable mixture of a quality not less than that specified in BS. 1452:1961, Grade C. Cast iron is the 'traditional' material for bath making and still claims large portion of sales.
PROTECTIVE COATING	**Fireclay** has a white coating applied before glazing to cover the buff coloured base material. **Vitreous China** does not require this preliminary treatment. The final coating of ceramic glaze is approximate to glass coating fused at high temperature. Can be transparent or coloured.	Usually referred to as porcelain enamel. Preliminary heat treatment of about 800°C (1470°F) then cleaned by blasting. Undercoat is then swilled, brushed or sprayed on, then dried and fused. Cover coat of white or coloured 'frit' or enamel is then sprayed on, dried and fused at temperature of 680°-780°C (1260°-1440°F).
IMPORTANT FEATURES	**Fireclay.** Suitable for very hard use being heavier and bulkier. Ideal for the larger appliances for hospitals. Life unlimited in normal use. (Semi-porous, therefore glazing is important). **Vitreous China.** Impervious to moisture even when unglazed. Strong material but comparatively light in weight. Eminently suitable for general use.	Strong and stable material which is very hard wearing. Entirely free from distortion during use. Surface has a high lustre which is extremely durable.
HYGIENE & CLEANING	Hygienic properties depend upon the glaze adherence which in the case of vitreous china is particularly good. Can be easily cleaned without damage to the surface.	Highly resistant to abrasion but correct cleaning powders should be used. If cleaned immediately after use, the task is an easy one.
WEIGHT	If support is required, vitreous china requires less support than fireclay ware. Ceramics are inclined to be rather heavy but depends upon the appliance selected.	Very heavy material which naturally produces a heavy appliance. Weight of typical baths varies between 90 kg and 113 kg with the maximum about 136 kg.
MISCELLANEOUS POINTS	High thermal capacity, particularly the larger article. The widest range of appliances to select from is made from ceramic materials.	Thermal capacity is quite high. Awkward and difficult to handle due to weight/bulk but stays in position without further fixing.

Sanitation: Appliances

STEEL VITREOUS ENAMELLED

First quality enamelling steel for pressing or deep drawing to required shape, with minimum of two coats of vitreous enamel (ground coat and cover coat). Fired at high temperature between 750°-900° (1380-1650°F). The basic steel characteristics are low carbon content and freedom from surface blemishes.

Vitreous enamel (sometimes known as porcelain enamel) is opaque glass (frit) fused to the metal forming a permanent bond. Enamels are specially formulated to suit each particular application and constitute electrical insulation. Produces a smooth even glossy finish in a variety of colours.

Low first coat (compared with stainless steel). Very durable in use, retaining good appearance throughout the life of the article due to the extreme hardness of vitreous enamel. Availability of attractive colours which are permanent and non-fading.

Surface very hygienic and cleaning easily carried out with any proprietary cleaner. Harsh abrasives should not be used. Surface will not harbour harmful bacteria. Regular cleaning should consist of wiping with damp cloth only.

Relatively light in weight. Typical weight of 1.0 × 0.5 m sink top is 14.5 kg. Bath at 1.8 m weighs approximately 41 kg and needs special cradle support (supplied).

Relatively low thermal capacity. Inclined to produce noise problem when in use. Severe blows can cause chipping of the enamel.

STAINLESS STEEL

The term stainless is given to highly alloyed steels containing large proportions of chromium. The type for sanitary appliances is known as 18/8 (minimum content 18% chromium and 8% nickel). Identified by its metallic appearance which can be 'mirror' or 'satin'. Satin has greater resistance to marking and scratching.

No protective coating necessary. Stainless steel has inherent corrosion resistance due to the hard, adherent and transparent oxide film which covers the surface of the metal and reforms instantaneously if the metal is scratched or abraded.

Absolute resistance to corrosion by solids or liquids with which it comes into contact. Exceptions: concentrated chlorinated bleach and certain acids not found in domestic circumstances. Life should be infinite providing deliberate damage is not caused.

Cleaning can be carried out with mild proprietary cleaner or one specifically designed for the care of stainless steel. Not affected by correctly used domestic cleaning materials but harsh abrasives must be avoided. Very hygienic.

Relatively light in weight, even lighter than vitreous enamelled steel. Typical weight of 1.0 × 0.5 m sink top is 10 kg (sink top sizes approximate).

Relatively low thermal capacity. Sound deadening material is applied to lessen the noise nuisance. Severe blows only dent this material.

PLASTIC MATERIALS

'Perspex' — acrylic sheet (polyethyl methacrylate) thermally formed, with high gloss appearance.
Polypropylene — polymerised propylene with smooth gloss appearance.
Glass Reinforced Plastic — resin (e.g. polyester) reinforced with glass fibres.
Nylon — synthetic polymeric amines with smooth appearance.

No protective coating is necessary and most plastic materials are self-coloured throughout, homogenous and free from ripples or blemishes. Glass reinforced plastics rely on a gel coat to protect the underlying fibres. 'Perspex' coloured ware is easy to match with ceramic coloured ware.

Details refer mainly to perspex appliances. Very tough material; does not chip; light to handle; coloured throughout; warm to the touch; slip resistant and resilient. Full range of standard colours can be supplied at no extra cost. Can be damaged by abrasion, lighted cigarettes and hot pans.

Very hygienic and not affected by many substances that attack metals. Clean with warm soapy water. Slight dulling of the surface can be removed by metal polish and scratches can be erased by use of abrasive and metal polish, if appliance is 'Perspex'.

Very light material. Baths made of 'Perspex' have a cradle to support and make bath rigid. Typical weight of bath is 11 kg to 14 kg without the cradle (approximate).

Good thermal insulator. Allowance must be made for expansion. Practically no noise problem as there may be with iron and steel.

SOAP DISPENSERS

Tilt type
for liquid soap

Sample capacities in litres
0.28, 0.38, 0.43, 0.57, 0.71

'York'
(360 ml)

Stainless steel

Toughened glass bowl

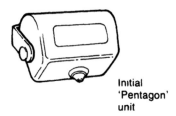

Initial 'Pentagon' unit

Special key required to remove nozzle for filling

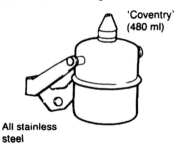

'Coventry' (480 ml)

All stainless steel

Soaps
Choice of industrial and toilet grades.
Choice of perfumes and colours
Examples of quantities supplied 4.5l tins, 22.7l and 205l drums

Plunger type
for liquid or cream soap

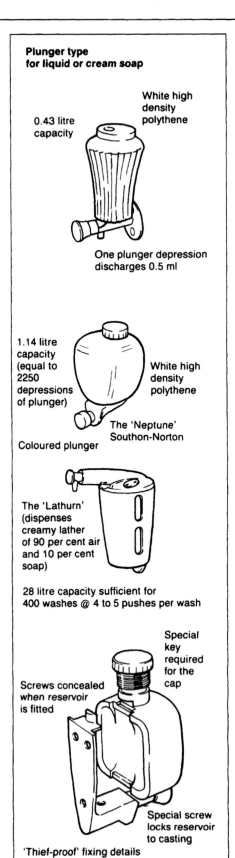

White high density polythene

0.43 litre capacity

One plunger depression discharges 0.5 ml

1.14 litre capacity (equal to 2250 depressions of plunger)

White high density polythene

The 'Neptune' Southon-Norton

Coloured plunger

The 'Lathurn' (dispenses creamy lather of 90 per cent air and 10 per cent soap)

28 litre capacity sufficient for 400 washes @ 4 to 5 pushes per wash

Special key required for the cap

Screws concealed when reservoir is fitted

Special screw locks reservoir to casting

'Thief-proof' fixing details of the 'Neptune'

Basin type

Can be used as a second matching tap to 'blended' water tap, or as a third tap when hot and cold water is supplied

'Econa'

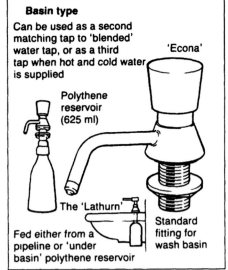

Polythene reservoir (625 ml)

The 'Lathurn'

Standard fitting for wash basin

Fed either from a pipeline or 'under basin' polythene reservoir

Wash fountain type

Porcelain enamelled soap reservoir

Can be fed from soap reservoir or by pipeline from central storage

Up to 6 plungers can be provided

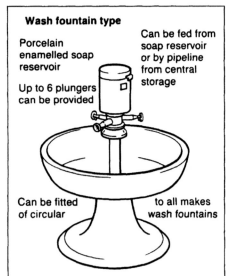

Can be fitted of circular

to all makes wash fountains

Miscellaneous

0.568 litre cap

Refill soap container 4.6 litre capacity

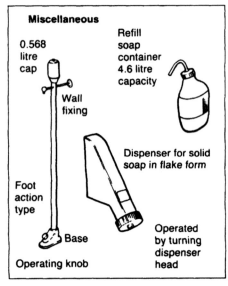

Wall fixing

Foot action type

Dispenser for solid soap in flake form

Operating knob

Base

Operated by turning dispenser head

SOAP DISPENSING SYSTEMS

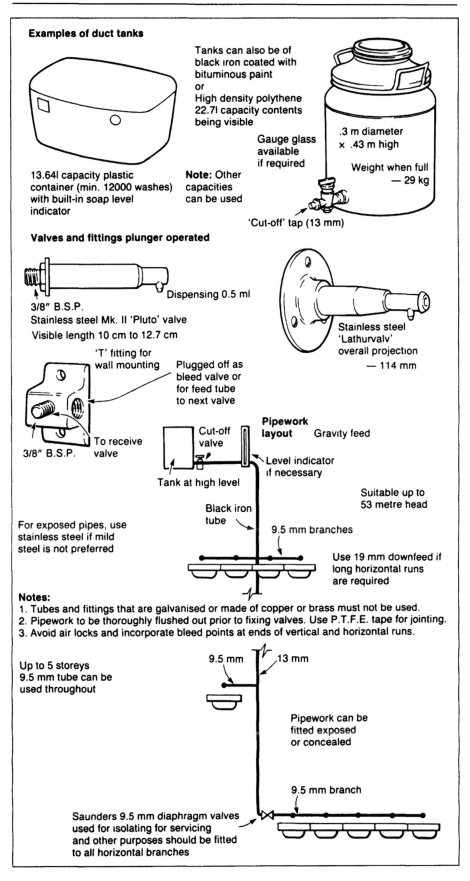

Examples of duct tanks

Tanks can also be of black iron coated with bituminous paint
or
High density polythene 22.7l capacity contents being visible

.3 m diameter × .43 m high

Weight when full — 29 kg

Gauge glass available if required

13.64l capacity plastic container (min. 12000 washes) with built-in soap level indicator

Note: Other capacities can be used

'Cut-off' tap (13 mm)

Valves and fittings plunger operated

Dispensing 0.5 ml

3/8" B.S.P.
Stainless steel Mk. II 'Pluto' valve
Visible length 10 cm to 12.7 cm

Stainless steel 'Lathurvalv' overall projection — 114 mm

'T' fitting for wall mounting

Plugged off as bleed valve or for feed tube to next valve

3/8" B.S.P.

To receive valve

Pipework layout Gravity feed

Cut-off valve

Level indicator if necessary

Tank at high level

Suitable up to 53 metre head

Black iron tube

9.5 mm branches

For exposed pipes, use stainless steel if mild steel is not preferred

Use 19 mm downfeed if long horizontal runs are required

Notes:
1. Tubes and fittings that are galvanised or made of copper or brass must not be used.
2. Pipework to be thoroughly flushed out prior to fixing valves. Use P.T.F.E. tape for jointing.
3. Avoid air locks and incorporate bleed points at ends of vertical and horizontal runs.

Up to 5 storeys 9.5 mm tube can be used throughout

9.5 mm

13 mm

Pipework can be fitted exposed or concealed

9.5 mm branch

Saunders 9.5 mm diaphragm valves used for isolating for servicing and other purposes should be fitted to all horizontal branches

1. They ensure a saving in running costs when compared with tablet soap, estimates varying between 50 per cent and 75 per cent.
2. They prevent waste inasmuch as soap cannot be left in the water to dissolve and cannot block the pipes.
3. They ration the amount of soap used and liquid soap cannot be left running.
4. They are easily serviced by an authorised person and are designed to prevent pilfering.
5. They eliminate the risk of infection as each user gets clean fresh soap.

13

TOWELS

Paper towel dispensers

Paper from the roll is dispensed in controlled lengths and delivered one sheet at a time Weight (empty 4.5 kg). Weight of roll 1.4 kg

Length of roll — 110 m providing 360 pieces

292 mm

248 mm

178 mm

Cutting edge

254 mm

Paper towel dispensers — Advance Services

250 mm

Continuous towel dispenser

Vitreous enamelled dispenser cabinet

Towels are of white huckaback

Individual paper towel dispenser

Towels obtainable in sleeves of 125 or in cartons of 4000

Folded — 100 mm
Opened — 305 mm

Semi-automatic Paper Master — Dispenses paper towel from a roll with mechanical check and timer for economic usage. The steel casing measures 290 mm by 240 mm by 180 mm. Rolls of Kraft paper are 110 m by 250 mm.

Interleaf Paper Master —
Dispenses interleaved paper towels measuring 350 mm by 270 mm by 100 mm. The sleeve contains 150 sheets. The unit has a steel casing and no moving parts.

Continuous towel dispenser — Huckaback 100 per cent cotton towel rolls measuring 40 m by 275 mm wide. Size of the unit is 460 mm by 380 mm by 230 mm. There is a timer mechanism to discourage waste, and a Yale-type cabinet lock.

Initial Towelflow Cabinets are installed and serviced free of charge and rolls of clean towels delivered regularly according to requirements for an inclusive rental.

Examples of dispensers

	Standard	Midget
Sizes of dispensers	457 mm high 381 mm wide 216 mm deep	330 mm high 381 mm wide 165 mm deep
Approximate towel size	280 × 41.2 m	280 × 13.7 m
No. of pulls (Lengths of clean towel)	230	100
Uses	Offices, factories, schools etc.	Executive suites, medical and dental surgeries

FLUSH FITTING TOWEL CABINET

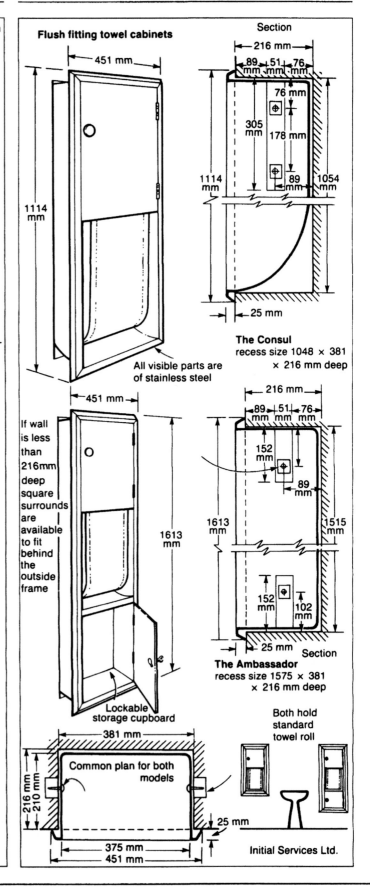

Flush fitting towel cabinets

451 mm

1114 mm

All visible parts are of stainless steel

Section

216 mm

89 mm 51 mm 76 mm

76 mm

305 mm

178 mm

1114 mm

89 mm

1054 mm

25 mm

The Consul
recess size 1048 × 381 × 216 mm deep

451 mm

If wall is less than 216mm deep square surrounds are available to fit behind the outside frame

1613 mm

Lockable storage cupboard

216 mm

89 mm 51 mm 76 mm

152 mm

89 mm

1613 mm

1515 mm

152 mm

102 mm

25 mm Section

The Ambassador
recess size 1575 × 381 × 216 mm deep

381 mm

216 mm
210 mm

Common plan for both models

375 mm

451 mm

25 mm

Both hold standard towel roll

Initial Services Ltd.

Sanitation: Equipment

ELECTRIC HAND DRYERS

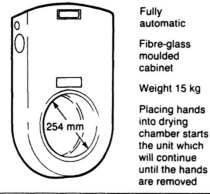

Electronically controlled wall model

Broughton

362 mm wide
629 mm high
267 mm deep

30-40 sec. operation

Heater 1.0 kW

Standby load
220 milliamps

254 mm

Fully automatic

Fibre-glass moulded cabinet

Weight 15 kg

Placing hands into drying chamber starts the unit which will continue until the hands are removed

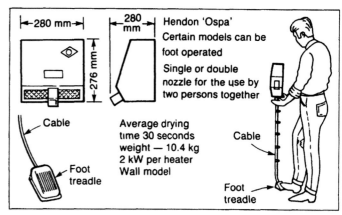

├─280 mm─┤   ├─280 mm─┤

276 mm

Cable

Foot treadle

Hendon 'Ospa'

Certain models can be foot operated

Single or double nozzle for the use by two persons together

Average drying time 30 seconds
weight — 10.4 kg
2 kW per heater
Wall model

Cable

Foot treadle

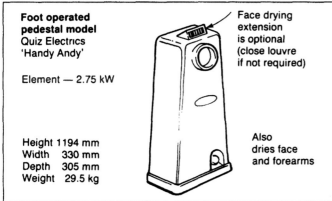

Push button

20 sec. time cycle

Paper towel

Wall model Coinamatic

Operates automatically when paper towel is pulled. A 60 watt preheater is on all the time, plus another 1500 watt when fan and main heater is in operation

406 mm high × 343 mm wide
× 241 mm deep

Contains 250 paper towels

Combines warm air and use of paper towel

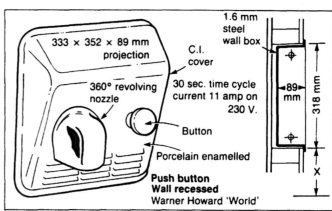

333 × 352 × 89 mm projection

360° revolving nozzle

C.I. cover

1.6 mm steel wall box

30 sec. time cycle current 11 amp on 230 V.

89 mm

318 mm

Button

Porcelain enamelled

X

**Push button
Wall recessed**
Warner Howard 'World'

For gentlemen   HT 'x' = 1092 mm
For ladies   HT 'x' = 1016 mm
Children 4-7 years   HT 'x' = 711 mm
Children 7-10 years HT 'x' = 813 mm

Some advantages claimed by the manufacturers of electric hand-dryers for their appliances.
1. Money saving by eliminating expenditure on towel replacement, laundry bills and the problem of unaccountable towel losses.
2. Labour saving by eliminating distribution and collection of clean and soiled towels. No disposal problems as for paper towels.
3. Improved hygiene by eliminating risk of infection in communal use of towels and provides more thorough drying action.
4. Disadvantage could be that capital costs must be faced up to, plus any maintenance that may become necessary.

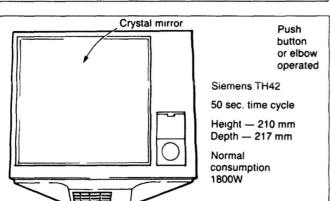

Foot operated pedestal model
Quiz Electrics 'Handy Andy'

Element — 2.75 kW

Height 1194 mm
Width   330 mm
Depth   305 mm
Weight 29.5 kg

Face drying extension is optional (close louvre if not required)

Also dries face and forearms

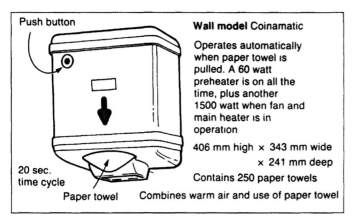

Crystal mirror

Push button or elbow operated

Siemens TH42

50 sec. time cycle

Height — 210 mm
Depth — 217 mm

Normal consumption 1800W

SMALL INCINERATORS

Electric-powered incinerators or water macerators are the most common form of unit used for sanitary towel disposal, although some gas-fired units can be found. Suitable installations for this type of appliance would be hospitals, offices, schools and communal dwellings.

Provision should be one unit per 25 communal dwellings, or one unit per 150 persons.

Gas

Gas units available include the Victor 352 Mk. II. By opening the load door, the gas ignites and burns for 10 minutes, after which it turns off automatically. The incinerator can be loaded at any time during the burning cycle, every time the loading door is opened, the incinerator will burn for a further 10 minutes. This unit is capable of burning up to 60 towels per hour.

Victor 352 Mk.II
Flue — heavy duty asbestos, 102 mm internal diameter to BS.835 (1984).
Capacity — combustion chamber capacity is 5.6 litres.
Weight — 15.9 kg.
Input — 10,000 btu/hr.

Electric

Operation of these units is 15 minute time cycle which commences on closing the access door. It will recommence the cycle if the access door is opened during the cycle. Various models allow flush- or surface-mounting. Units operate by the emission of a superheated air jet from a central nozzle which ignites the waste. The refractory material combustion chamber has a cast iron hearth beneath. The ash draw should be emptied regularly.

Water Macerators

This is the most popular type of sanitary towel disposal unit. It is available in both floor-mounted and wall-hung models, and each unit can serve up to 250 staff.

Sanimatic 125
Drainage — by 38 mm diameter outlet with 75 mm deep seal 'S' trap; pipe run direct to discharge stack with minimum number of bends, and 90 deg. bends are not recommended.

Water Supply — to be via separate down service from the building supply or from separate break tank. Minimum 1.5 m above unit: supply pipe —

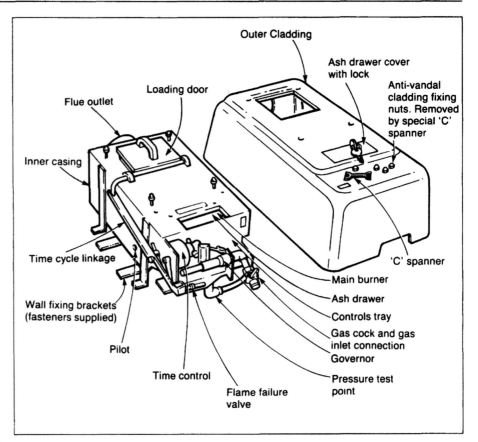

Outer Cladding
Ash drawer cover with lock
Anti-vandal cladding fixing nuts. Removed by special 'C' spanner
Loading door
Flue outlet
Inner casing
Time cycle linkage
Wall fixing brackets (fasteners supplied)
Pilot
Time control
Flame failure valve
'C' spanner
Main burner
Ash drawer
Controls tray
Gas cock and gas inlet connection
Governor
Pressure test point

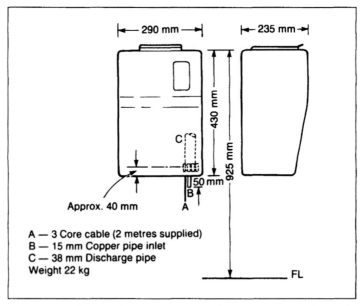

|← 290 mm →| |← 235 mm →|

430 mm
925 mm
50 mm
Approx. 40 mm
A
B
C

A — 3 Core cable (2 metres supplied)
B — 15 mm Copper pipe inlet
C — 38 mm Discharge pipe
Weight 22 kg
FL

35 mm diameter copper with full bore lockshield gate valve at the inlet position. Minimum flow rate required is 8 litres per minute. Where the static head exceeds 3 m (4.4 psi/0.3 kg/cm³) a pressure reducing valve or

alternative solenoid valve must be fitted.

Electrical — Motor is 0.37 kW, 240V single-phase, 50 Hz, operating at 1500 rpm.

RECESSED MODEL

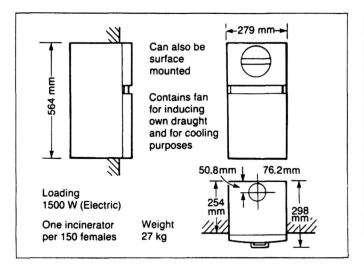

Can also be surface mounted

Contains fan for inducing own draught and for cooling purposes

564 mm

←279 mm→

50.8mm ↕ 76.2mm
254 mm
298 mm

Loading
1500 W (Electric)

One incinerator per 150 females

Weight 27 kg

SEALED CONTAINER DISPOSER

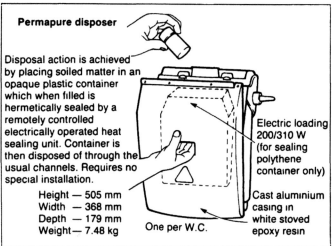

Permapure disposer

Disposal action is achieved by placing soiled matter in an opaque plastic container which when filled is hermetically sealed by a remotely controlled electrically operated heat sealing unit. Container is then disposed of through the usual channels. Requires no special installation.

Height — 505 mm
Width — 368 mm
Depth — 179 mm
Weight— 7.48 kg

One per W.C.

Electric loading 200/310 W (for sealing polythene container only)

Cast aluminium casing in white stoved epoxy resin

BULK DISPOSAL

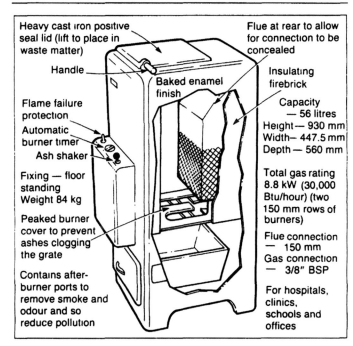

Heavy cast iron positive seal lid (lift to place in waste matter)

Handle

Flame failure protection

Automatic burner timer

Ash shaker

Fixing — floor standing
Weight 84 kg

Peaked burner cover to prevent ashes clogging the grate

Contains after-burner ports to remove smoke and odour and so reduce pollution

Baked enamel finish

Flue at rear to allow for connection to be concealed

Insulating firebrick

Capacity — 56 litres
Height— 930 mm
Width— 447.5 mm
Depth — 560 mm

Total gas rating 8.8 kW (30,000 Btu/hour) (two 150 mm rows of burners)

Flue connection — 150 mm
Gas connection — 3/8" BSP

For hospitals, clinics, schools and offices

Notes
Choice of site is very important. Incinerators must be connected to an efficient flue. Therefore an exterior wall may be preferable consisting of non-combustible material capable of supporting the unit. For convenience of loading, site close as possible to the source of waste. For sanitary towel disposal they should be installed in a toilet. Units may be used for disposal of small surgical dressings etc. in which case they may be installed in clinics or other similar locations. Incinerators can also be used for the destruction of legal and confidential documents.

DISPOSERS WITH SHREDDING ACTION

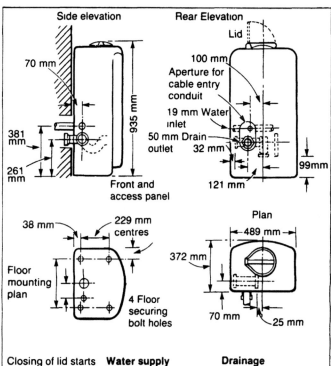

Side elevation

Rear Elevation

Lid

70 mm
935 mm
381 mm
261 mm

Front and access panel

100 mm
Aperture for cable entry conduit
19 mm Water inlet
50 mm Drain outlet 32 mm
121 mm
99mm

38 mm

229 mm centres

Floor mounting plan

4 Floor securing bolt holes

Plan
←489 mm→
372 mm
70 mm 25 mm

Closing of lid starts fixed time cycle and towels are disintegrated and flushed away through the drainage system. If lid is lifted during cycle, shredding action stops and re-commences when lid is closed.

Water supply
Minimum flow of 20 litres/second must be maintained. Size of inlet connection 19 mm

Water supply
Minimum head of water 1.2 m from own cistern supply. If head less than 3 m use 32 mm supply pipe

Drainage
Unit is supplied with 75 mm deep seal trap. Waste pipe to be 50 mm. Connect to nearest soil drain

Electrical data
Motor 2.24 kW
Volts 400/440 AC
50 Hz 3 phase
maximum load — 4.57 amps
Revs per second — 47.5

INCINERATORS AND NATURAL DRAUGHT

Notes concerning flue design
1. Use independent flue for each incinerator and allow the manufacturers recommended distance beneath the appliance.
2. Minimum vertical height of flue measured from top of appliance = 2.7 m.
3. Use vertical length of metal flue pipe (0.6 to 0.9 m) directly onto the appliance, then a change of direction can be made if required.
4. Bends and horizontal runs reduce the 'pull', so for preference the appliance should be fixed on or near an external wall or duct.
5. Horizontal runs should be not more than 1/3 of the vertical lift with a maximum of 3 m. Use obtuse bends (with access for periodic cleaning). Rule of thumb allowances — allow 0.3 metre extra height per bend, and ditto for each .15 m horizontal run.
6. Draught diverters must not be used nor updraught restricted.
7. When siting terminals, avoid down-draught positions, depressive areas and overshadowing by high walls, buildings, etc. Always terminate above eaves or parapet wall and well away from windows.
8. Horizontal terminal positions may be used in certain circumstances.
9. Terminals should have cowl fitted. If vertical — open type with cone cap.
10. Natural ventilation to room containing the incinerator is necessary to provide air for combustion. If the room is to be mechanically ventilated, natural draught flues must not be used unless a separate air supply is provided.

Wandsworth 'Bunnie' +
Each incinerator has a built-in fan unit to induce smoke and fumes from combustion chamber to pass up the flue. A relay will switch on all the fans in the incinerators not in use, once any machine has been put to work. For more than 6 machines, or if long horizontal flues are involved, an extracter fan at roof level should be incorporated and wired to operate if any machine is used.

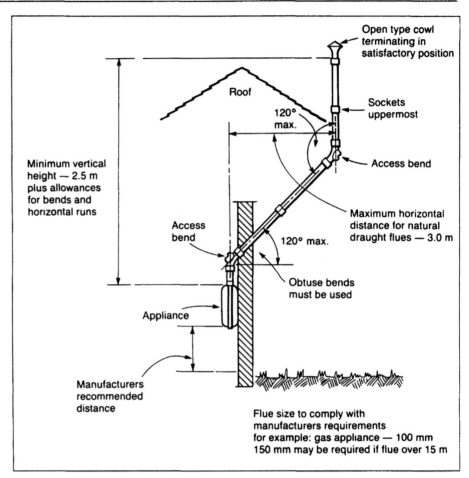

Open type cowl terminating in satisfactory position
Sockets uppermost
Access bend
Maximum horizontal distance for natural draught flues — 3.0 m
Obtuse bends must be used
120° max.
120° max.
Roof
Minimum vertical height — 2.5 m plus allowances for bends and horizontal runs
Access bend
Appliance
Manufacturers recommended distance

Flue size to comply with manufacturers requirements for example: gas appliance — 100 mm 150 mm may be required if flue over 15 m

INCINERATOR SYSTEMS

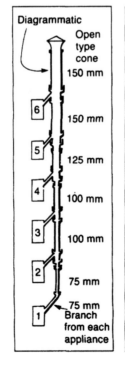

Diagrammatic

Open type cone
150 mm
150 mm
125 mm
100 mm
100 mm
75 mm
75 mm Branch from each appliance

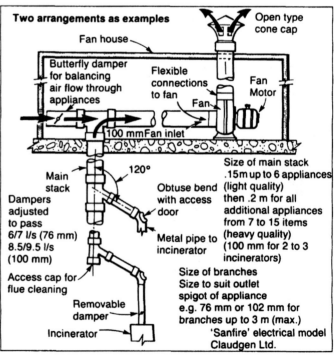

Two arrangements as examples

Fan house
Open type cone cap
Butterfly damper for balancing air flow through appliances
Flexible connections to fan
Fan Motor
Fan
100 mm Fan inlet
Main stack
120°
Dampers adjusted to pass 6/7 l/s (76 mm) 8.5/9.5 l/s (100 mm)
Obtuse bend with access door
Metal pipe to incinerator
Access cap for flue cleaning
Removable damper
Incinerator

Size of main stack .15 m up to 6 appliances (light quality) then .2 m for all additional appliances from 7 to 15 items (heavy quality) (100 mm for 2 to 3 incinerators)

Size of branches Size to suit outlet spigot of appliance e.g. 76 mm or 102 mm for branches up to 3 m (max.)
'Sanfire' electrical model Claudgen Ltd.

Sanitation: Equipment

DISPOSAL BY MECHANICAL SYSTEMS

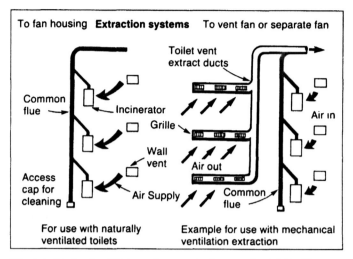

To fan housing **Extraction systems** To vent fan or separate fan

- Common flue
- Incinerator
- Grille
- Wall vent
- Air out
- Air Supply
- Access cap for cleaning
- Air Supply
- Toilet vent extract ducts
- Air in
- Common flue

For use with naturally ventilated toilets

Example for use with mechanical ventilation extraction

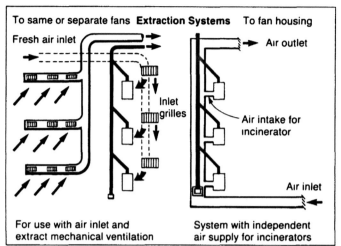

To same or separate fans **Extraction Systems** To fan housing

- Fresh air inlet
- Air outlet
- Inlet grilles
- Air intake for incinerator
- Air inlet

For use with air inlet and extract mechanical ventilation

System with independent air supply for incinerators

Mechanical extraction systems may be employed for the following reasons:
1. Where natural draught fails or is seen to be impracticable (single appliances can have fan fitted in flue duct).
2. When incinerators are fitted in toilets subjected to ventilation by mechanical air extraction (depressurised), the use of natural draught is not possible unless a separate air supply to the incinerators is provided.
3. Where common flues are employed serving a number of incinerators installed one above the other in multi-storey buildings. Such a system avoids the multiplicity of separate flues, saving space and reducing building costs.

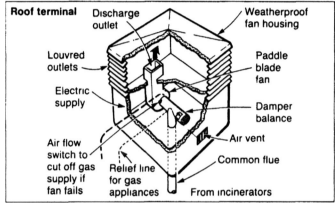

Roof terminal
- Discharge outlet
- Weatherproof fan housing
- Louvred outlets
- Paddle blade fan
- Electric supply
- Damper balance
- Air flow switch to cut off gas supply if fan fails
- Air vent
- Common flue
- Relief line for gas appliances
- From incinerators

SHREDDING SYSTEM

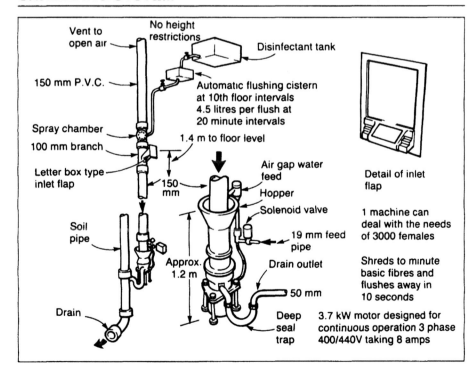

- Vent to open air
- No height restrictions
- Disinfectant tank
- 150 mm P.V.C.
- Automatic flushing cistern at 10th floor intervals 4.5 litres per flush at 20 minute intervals
- Spray chamber
- 100 mm branch
- 1.4 m to floor level
- Letter box type inlet flap
- 150 mm
- Air gap water feed
- Hopper
- Solenoid valve
- Soil pipe
- 19 mm feed pipe
- Approx. 1.2 m
- Drain outlet
- Drain
- 50 mm
- Deep seal trap

Detail of inlet flap

1 machine can deal with the needs of 3000 females

Shreds to minute basic fibres and flushes away in 10 seconds

3.7 kW motor designed for continuous operation 3 phase 400/440V taking 8 amps

BUILDING REGULATIONS REMINDERS

Building Regulations Reminders
In the Approved Document J to the 1985 Building Regulations, Section 2 relates to gas burning appliances with a rated input up to 60kW, and gives provisions for:
a) cooking appliances
b) balance-flued appliances
c) decorative log and other solid fuel fire effect gas appliances (decorative appliances)
d) other individual, natural draught, open-flued appliances.
Any other appliances should be installed in accordance with the relevant recommendations of BS.5440 Code of Practice for flues and air supply for gas appliances of rated input not exceeding 60 kW (first- and second-family gases) Part 1:1978 Flues, and Part 2:1976 Air Supply.

DOMESTIC FOOD WASTE DISPOSERS

Purpose
These units are designed to dispose of organic food waste electro-mechanically within seconds and flush the residue into the drain. The unit should dispose of all types of food waste, preloaded or fed continuously, but metal, rags and plastic objects should not be placed in the disposer. Food waste is therefore hygienically 'destroyed' at the sink.

Operation
Turn on cold water switch on disposer and feed water into the unit (unless preloaded). Waste falls onto high speed rotor and is flung against the stationary cutting ring with tremendous centrifugal force, shredding the waste into tiny particles. This part liquified waste, filters through the rotor into the waste pipe with the flow of water, the latter also keeping the machine clean and free from unpleasant odour. A thermal overload device cuts off the power in the event of jamming. After-sales service is not normally needed.

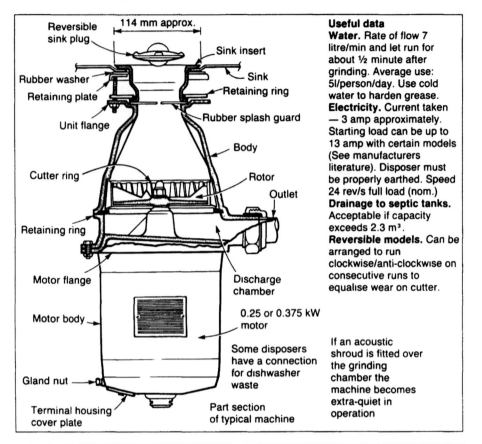

Part section of typical machine

Useful data
Water. Rate of flow 7 litre/min and let run for about ½ minute after grinding. Average use: 5l/person/day. Use cold water to harden grease.

Electricity. Current taken — 3 amp approximately. Starting load can be up to 13 amp with certain models (See manufacturers literature). Disposer must be properly earthed. Speed 24 rev/s full load (nom.)

Drainage to septic tanks. Acceptable if capacity exceeds 2.3 m³.

Reversible models. Can be arranged to run clockwise/anti-clockwise on consecutive runs to equalise wear on cutter.

Some disposers have a connection for dishwasher waste

If an acoustic shroud is fitted over the grinding chamber the machine becomes extra-quiet in operation

Sink connections

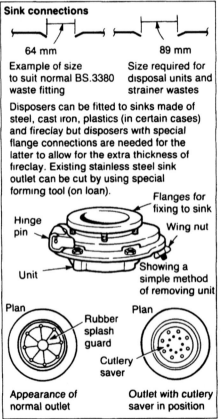

64 mm
Example of size to suit normal BS.3380 waste fitting

89 mm
Size required for disposal units and strainer wastes

Disposers can be fitted to sinks made of steel, cast iron, plastics (in certain cases) and fireclay but disposers with special flange connections are needed for the latter to allow for the extra thickness of fireclay. Existing stainless steel sink outlet can be cut by using special forming tool (on loan).

Appearance of normal outlet

Outlet with cutlery saver in position

Waste connections
Note: Domestic disposers have no integral trap

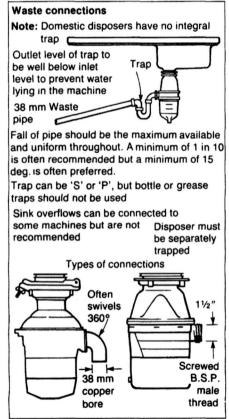

Outlet level of trap to be well below inlet level to prevent water lying in the machine

38 mm Waste pipe

Fall of pipe should be the maximum available and uniform throughout. A minimum of 1 in 10 is often recommended but a minimum of 15 deg. is often preferred.

Trap can be 'S' or 'P', but bottle or grease traps should not be used

Sink overflows can be connected to some machines but are not recommended

Disposer must be separately trapped

Types of connections

Often swivels 360°

38 mm copper bore

1½"

Screwed B.S.P. male thread

COMMERCIAL DISPOSERS

Electrical

Reversing controller (alternates direction of impellor)

Fused double pole spur box flush switched with pilot lamp

On/off switch

Controls

Disposers have thermal overload devices in case of jamming. Reset buttons are either on the machine or the control unit. Procedure:- switch off, remove stoppage, press button until it clicks, switch on

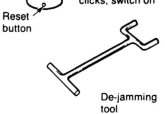

Reset button

De-jamming tool

Selection of models available
1. Catering models with a range of kW motors from 0.375, 0.56, 1.12, 2.24, 3.73, 5.6 to 7.5.
2. For disposal of putrescible waste.
3. For disposal of maternity waste.
4. For disposal of small specimens from laboratories.
5. For sanitary towel or similar waste disposal.

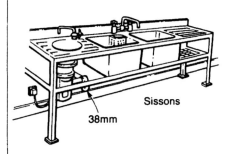

Sissons

38mm

Catering sink with 'Whirl-a-waste' disposer fitted to drainer. Can be bolted or welded to dishwasher or preparation table. Capacity of 206 kg of normal mixed food waste.

Intended for the disposal of sanitary towels and tissues by disintegration and flushing away through the drainage system. Closing lid starts 15 sec. cycle.

Water supply 19 mm to give minimum flow of 20 l/minute

Waste size 50 mm

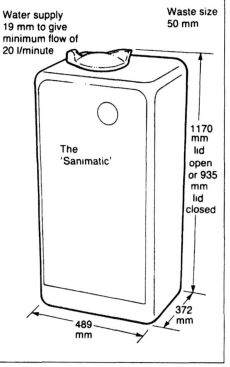

The 'Sanimatic'

1170 mm lid open or 935 mm lid closed

372 mm

489 mm

Discharge of wastes

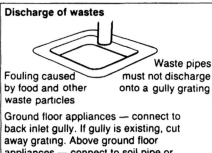

Fouling caused by food and other waste particles

Waste pipes must not discharge onto a gully grating

Ground floor appliances — connect to back inlet gully. If gully is existing, cut away grating. Above ground floor appliances — connect to soil pipe or waste pipe.

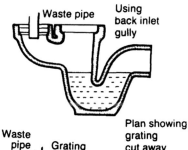

Waste pipe

Using back inlet gully

Waste pipe   Grating

Plan showing grating cut away

Miscellaneous

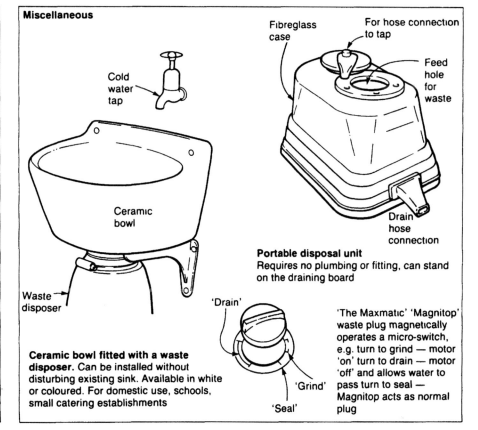

Fibreglass case

For hose connection to tap

Feed hole for waste

Cold water tap

Ceramic bowl

Drain hose connection

Waste disposer

Portable disposal unit
Requires no plumbing or fitting, can stand on the draining board

Ceramic bowl fitted with a waste disposer. Can be installed without disturbing existing sink. Available in white or coloured. For domestic use, schools, small catering establishments

'Drain'

'Grind'

'Seal'

'The Maxmatic' 'Magnitop' waste plug magnetically operates a micro-switch, e.g. turn to grind — motor 'on' turn to drain — motor 'off' and allows water to pass turn to seal — Magnitop acts as normal plug

PART G2 — BATHROOMS

A bathroom shall be provided containing a fixed bath or shower/bath, and there shall be a suitable installation for the provision of hot and cold water to the bath or shower/bath. This requirement applies only to dwellings, and replaces Section 28 of the Building Act 1984.

PART G4 — SANITARY CONVENIENCES

Sufficient sanitary conveniences will be provided, which shall be:
a) in rooms separate from places where food is stored or prepared,
b) designed and installed so as to allow effective cleaning.

(This section replaces Section 26 of the Building Act 1984. Local authorities have a duty to see that there are sanitary conveniences in factories, workshops, and work places (Section 65) and that a W.C. is constantly supplied with water for flushing and is protected from frost (Section 51 of the Public Health Act 1936). The Health and Safety Executive, or local authority, has a duty to see that there are sanitary conveniences and washing facilities provided in premises to which the Offices, Shops and Railway Premises Act 1963 applies. The Food Hygiene (General) Regulations 1970 apply to premises used for the purposes of a food business.)

APPROVED DOCUMENT G4

Sanitary Conveniences lays down the minimum 'acceptable' level of performance as follows:

To reduce the risks to the health of the persons in the buildings, closets should be provided which are
a) in sufficient number and of the appropriate type for the sex and age of persons using the building, and
b) be sited, designed and installed so as not to be prejudicial to health.

Sanitation: Legislation

PUBLIC HEALTH ACTS 1936 AND 1961

Section — Sanitary Conveniences for Buildings

43 1936
Plans of new buildings to show sufficient and satisfactory closet accommodation. Waterclosets if sufficient water supply and sewer available, otherwise earth closets may be approved. Separate for sexes if factory or workplace and both sexes are employed.

44 1936 as amended by 21 1961
If a building, or any part of a building occupied as a separate dwelling is without sufficient closet accommodation or the existing closets so defective as to require reconstruction, L.A. shall require additional or substitute closets as necessary. (Does not apply to factory or workplace).

45 1936
If closets provided for a building are in such a state as to be prejudicial to health or a nuisance but capable of being put into satisfactory condition without reconstruction, L.A. shall by notice require necessary works or cleansing.

46 1936
Sanitary conveniences to be provided for work places. Related to number of persons employed, and if both sexes employed, to be separate for the sexes.

47 1936
Where sufficient water supply and sewer available, by waterclosets, either by owner or by L.A. If by owner — he is entitled to recover half cost from L.A. If by L.A. — entitled to recover half cost from owner. If owner proposes conversion without notice, L.A. may pay up to half cost.

Supplementary provisions

48 1936 as amended 16 1961
L.A. or M.O.H. or P.H.I., on reasonable grounds, may examine sanitary convenience believed to be in such condition as to be prejudicial to health or a nuisance. If satisfactory, L.A. to reinstate.

49 1936
Rooms over closets, other than waterclosets or earthclosets not to be used as living, sleeping or workplaces.

51 1936
Occupier required to keep watercloset supplied with flushing water and protected against frost — or earthcloset supplied with deodorising material.

52 1936
If sanitary convenience used by two or more families — penalty for injuring or improperly fouling, or for wilfully or negligently causing obstruction in drain from convenience.

17 1961
If W.C. stopped up, M.O.H. or P.H.I. can serve 48 hour notice on owner or occupier to remedy.

20 1961
If W.C. is so constructed or repaired as to be prejudicial to health or a nuisance, person who undertook or executed work liable to penalty unless he shows could not have been avoided by use of reasonable care.

22 1961
L.A. may, at request of owner or occupier, cleanse or repair watercloset or sink and make reasonable charge.

33 1961
Plans for a house, or part of a building to be occupied as a dwelling must show provision for fixed bath or shower with hot and cold water.

Public sanitary conveniences

87 1936
L.A. may provide in proper and convenient situations and may make Byelaws to control their use.

88 1936
Prohibits the erection of public sanitary conveniences accessible from the street, without consent of L.A.

89 1936
L.A. may require owner or occupier of inn, public house, beer house, refreshment house or place of public entertainment to provide sanitary conveniences for use of persons frequenting the premises.

80 1961
Defines refreshment house in above Section as "any building in which food and drink is sold to and consumed by the public".

Note: L.A. — local authority. M.O.H. — Medical Officer of Health. P.H.I. — Public Health Inspector.

Factories, Shops, Offices and

FACTORIES ACT 1961 (For definition of factory see Section 175 p.110-113)

Part 1 Health (General Requirements)
Section 1 Cleanliness.
(1) Every factory shall be kept in a clean state and free from effluvia arising from any drain, sanitary convenience or nuisance.

Section 7 Sanitary Conveniences. (1) Sufficient and suitable sanitary conveniences for persons employed to be provided, maintained and lighted. If both sexes are employed, to be separate for each sex, unless the only persons employed are members of the same family dwelling there. (2) Minister of Labour has power to make Regulations (Sanitary Accommodation Regulations 1938 continue in force by virtue of Section 183 and Sixth Schedule of Factories Act 1961 (ref. S.R. & O. 1938 No. 611).
The above Regulations provide 'inter alia' that sanitary conveniences shall be:
(5) Sufficiently ventilated, and shall not communicate with any workroom except through open air or intervening ventilated space, (6) (except in the case of urinals), under cover, partitioned off to secure privacy, have proper door and fastenings. Urinals to be placed or screened so as not to be visible from other parts of the factory where persons work or pass, (7) conveniently accessible to persons employed, (8) if both sexes are employed (a) so placed or screened that interior not visible even when doors are open, from any place where persons of other sex work or pass, (b) approaches to be separate if conveniences are adjoining, (c) conveniences for each sex to be indicated by suitable notice.

Part III Welfare (General Provisions)

Section 57 Supply of Drinking Water. (1) Adequate supply of wholesome drinking water shall be provided and maintained at suitable points, conveniently accessible to all persons employed. If not from a public main, to be from other source approved in writing by District Council.

Section 58 Washing Facilities. (1) Facilities for washing, including a supply of clean, running, hot and cold or warm water, and soap and clean towels or other suitable means of cleaning or drying shall be provided and maintained. (2 and 3) Authority for Minister of Labour to make Regulations regarding provision and regarding exemptions.

The latter has been done in Washing Facilities (Running Water) Exemption Regulations 1960. These remain in force by virtue of Section 183 and Sixth Schedule of Factories Act 1961 (Ref. S.I. 1960 No. 1029).

Section 59 Accommodation for Clothing. (1) Suitable accommodation for clothing not worn during working hours shall be provided and maintained, and arrangements made for drying such clothing. (2) Authority for Minister of Labour to make Regulations as to provision and exemptions from requirements.

PUBLIC HEALTH ACT 1936

(For definition of workplace see Section 343 p.211)

Section 46 Sanitary Conveniences in Workplaces. (1) Every 'workplace' to be provided with sufficient and satisfactory sanitary conveniences having regard to number of persons employed in or in attendance at the building. If both sexes are employed, separate accommodation required for each sex unless local authority is satisfied that in any particular case it is unnecessary.

Sanitation: Legislation

OFFICES, SHOPS AND RAILWAY PREMISES ACT 1963

Section 1. Premises to which this Act applies. Defines premises to which Act is applicable.

Sections 2 and 3 Exceptions. i.e. Cases in which the Act does not apply, e.g. only certain relatives employed, etc.

Health, Safety and Welfare of Employees (General Provisions)

Section 4 Cleanliness. All premises to which this Act applies, and all furniture, furnishings and fittings to be kept in a clean state. Minister of Labour may by Regulations prescribe requirements.

Section 9 Sanitary Conveniences. (1) Suitable and sufficient sanitary conveniences conveniently accessible to persons employed shall be provided. Arrangements may be made for sharing accommodation in appropriate cases (see Regulation 4 Statutory Instrument 1964 No. 966). (2) Shall be kept clean, properly maintained and effectively lighted and ventilated. (3) Authority for Minister of Labour to make Regulations prescribing standards (Sanitary Convenience Regulations 1964 Statutory Instrument 1964 No. 966). These Regulations provide 'inter alia' that:

Regulations 6. (a) No sanitary convenience shall be situated in any room in which any person is employed to work (other than a lavatory attendant). (b) No water closet or chemical closet, no accommodation in which a urinal is provided and no accommodation containing a W.C. or chemical closet, which in either case, is not wholly enclosed, shall be so situated that access to it is obtained directly from any room in which any person is employed to work. (c) If (b) is not practicable and the sanitary convenience was installed before 25th June 1964, mechanical means of ventilation to open air shall be provided and kept in operation while any person is employed in any room from which access is obtained. (d) Enclosed space between sanitary convenience and any room in which any person is employed shall be provided with effective means of ventilation.

Regulation 7. (a) All accommodation (W.C., chemical closet and urinal) to be covered to ensure protection from weather for persons using same. (b) To be sufficiently enclosed to ensure privacy and to be fitted with suitable door and fastenings. (c) Urinals to be positioned or screened so as not to be visible from outside the accommodation where the urinal is situated.

Regulation 8. Separate accommodation to be clearly marked to show the sex of persons for which it is provided.

Regulation 9. Suitable and effective means for disposal of sanitary dressings shall be provided if number of females employed exceeds ten. Any means so provided to be maintained in proper condition, and, if bins are used, contents shall be disposed of at suitable intervals.

Section 46. Gives power to enforcing Authority to grant exemptions from Section 9 in appropriate circumstances.

Section 10 Washing Facilities. (1) Suitable and sufficient washing facilities including clean, running hot and cold or warm water to be provided, together with soap and clean towels or other suitable means of cleaning or drying. Arrangements may be made for sharing accommodation in appropriate cases (see Regulation 4 Statutory Instrument 1964 No. 965). (2) Place to be kept clean, in orderly condition and effectively lighted. (3) Authority for Minister of Labour to make Regulations prescribing standards. (Washing Facilities Regulations 1964 Statutory Instrument 1964 No. 965). These Regulations provide 'inter alia' that:

Regulation 6. Accommodation for washing facilities shall be covered and enclosed sufficiently to ensure protection from the weather for persons using the facilities.

Regulation 7. Effective provision to be made as far as is reasonably practicable for venting rooms in which facilities are situated.

Regulation 8. Separate accommodation to be clearly marked to show the sex of persons for which it is provided.

Section 11 Supply of Drinking Water. An adequate supply of wholesome drinking water to be provided and maintained at suitable places conveniently accessible to all persons employed to work in the premises.

Section 12 Accommodation for Clothing. Similar to requirements of the Factories Act 1961 Section 59 (see Detail No. 19) with provision for the Minister to prescribe requirements by Regulations.

Section 45. Minister may, by order, grant exemptions from provision of Sections 9 and 10 in any class of premises.

Schools, Construction, Caravan Sites

SCHOOL PREMISES, CONSTRUCTION AND CARAVAN SITES

The Education (School Premises) Regulations 1981. Statutory Instrument 909

Made by S.O.S. for Education and Science under Section 10 of the Education Act 1944 (Statutory Instrument 1972 No. 2051)

Note: These Regulations deal separately with primary schools (Part II) secondary schools (Part III) nursery schools and nursery classes (Part IV) special schools (Part V) boarding accommodation (Part VI).

Regulations 7, 19, 31, 34 Storage of Children's Outdoor Clothing etc. Sufficient and suitable facilities for storing and drying pupils' outdoor clothing and for storing pupils' other belongings shall be provided.

Regulations 14, 27, 35, 40, 44 Accommodation for Meals. If the meals are cooked off the school premises, sufficient accommodation for receiving and serving out the meals and for washing up thereafter shall be provided.

Regulation 55 Water Supply. (1) Sufficient and wholesome water supply shall be provided for the premises. (2) The school shall be connected to piped water supply where reasonably practicable. (3) Where water under pressure is available, running water shall be laid on to various sanitary appliances. (4) Adequate warmed water shall be available at every washbasin, and water with emission temperature between 38 and 43.5°C at all baths and showers.

Regulations 8, 9, 11, 20, 21, 23, 24, 32, 39, 43, 57 Washing and Sanitary Accommodation. (1) Shall be soundly designed and constructed, reasonably accessible to the persons for whose use it was provided. (2) All urinals (stalls) to be of glazed material and fitted with flushing apparatus. (3) Each closet to have door and partition so constructed as to secure privacy. (4) Where water closets are provided each shall be capable of being flushed separately. (5) Surfaces of floors, including bath and shower compartments, and walls up to a minimum height of 1.8m shall be finished with a material which resists penetration of water and which can be easily cleaned.

Regulation 58 Drainage and Sewage Disposal. (1) Water closets shall be provided where a system of public sewers and a constant supply of water under pressure are available. (2) If no public sewers are available but a supply of water under pressure is, then similarly, water closets are to be provided. (3) Where public sewers are available but no water supply under pressure, drainage (which in this case includes wastes from urinals) is to be to sewer, and closets to be either earth or chemical. (4) In other cases, earth or chemical closets to be provided.

Health and Safety at Work Act 1974 Stat. Instr. 1966 No. 95

Regulation 11 Shelters and Accommodation for Clothing and for taking meals. (c) adequate and suitable accommodation affording protection from the weather, for taking meals shall be provided, plus facilities for boiling water. If contractor has more than 10 persons on a site and heated food is not available there shall be adequate facilities for heating food. All such accommodation shall be kept in a clean and orderly condition. (d) adequate supply of wholesome drinking water, at convenient points and clearly marked, shall be provided.

Regulation 12 Washing Facilities. In brief, contractor shall provide adequate and suitable facilities for washing. Actual requirements depend upon the number employed and completion time. Mention is made of providing adequate troughs, basins or buckets having in every case a smooth impervious internal surface, adequate and suitable means of cleaning and drying being soap and towels or other means, as the case may require, and a sufficient supply of hot and cold or warm water. If lead compound or other poisonous substance is used, nail brushes are to be provided. All to be kept in clean and orderly condition.

Regulation 14 Other Requirements as to Sanitary Conveniences. Includes... to be sufficiently ventilated and not to communicate with any workroom or messroom except through the open air or intervening ventilated space; to be undercover (except urinals), partitioned off as to secure privacy, have proper door and fastenings. Urinals to be effectively screened. To be conveniently accessible, maintained, kept clean, lit, etc.

Caravan Sites and Control of Development Act 1960

Section 5. Explains power of local authority to attach conditions to site licences.

Section 5(1)(f). For securing that adequate sanitary facilities, and such other facilities, services or equipment as may be specified, are provided and properly maintained.

Section 5(6). Minister of Housing and Local Government may specify model standards for purpose of this section.

Model Standards — Caravan Sites and Control of the Development Act 1960. Water supply to be to B.S. CP310 Water Supply. Provision to be made for foul drainage to public sewer, septic tanks or cesspool. For caravans without facilities, communal blocks should be provided.

Note: Local authority model standards apply to suit local conditions.

Sanitation: Legislation

FOOD PREMISES: FOOD HYGIENE (GENERAL) REGULATIONS 1970

Made jointly by Ministry of Agriculture, Fisheries and Food and S. of S. of Social Services under Section 13 and 123 Food Act 1984.

Regulation 15 Cisterns for supplying water to food rooms. Such cisterns shall not supply water to a sanitary convenience, except through an efficient flushing cistern or other flushing apparatus equally efficient and suitable for the prevention of contamination of water supplies.

Regulation 16 Sanitary Conveniences. (1)(a) Shall be kept clean and in efficient order, (1)(b) be so placed that no offensive odour therefrom can penetrate into any food room, (2) a room or other place which contains a sanitary convenience shall be suitably and sufficiently lighted and ventilated and kept clean, (3) a room containing a sanitary convenience not to be used as a food room, (4) if a food room communicates directly with a room or other place containing a sanitary convenience, it is not to be used for handling open food, (5) there shall be fixed and maintained in a prominent and suitable position near every sanitary convenience a clearly legible notice requesting food handlers to wash their hands after use.

Regulation 17 Water Supply to be provided. (1) A supply of water sufficient in quantity shall be provided. (2) Water to (a) be clean and wholesome; be constant if reasonably practicable, and to be in accordance with good practice.

Regulation 18 Wash hand basins to be provided. (1) Suitable and sufficient wash hand basins for use by all persons handling food to be placed in convenient and accessible position, (2) Every basin to have adequate supply of hot and cold water or 'blended' water; if no open food handled cold will suffice, (3) adequate supply of soap or other suitable detergent, nail brushes, clean towels or other suitable drying facilities at each basin, (4) basins to be kept clean and in good condition, (5) these facilities are only to be used for personal cleanliness of the user.

Regulation 20 Accommodation for Clothing, etc. (1) Suitable and sufficient accommodation for outdoor or other clothing and footwear not worn during working hours by persons handling food shall be provided in all food premises where open food is handled; clothing and footwear shall not be kept elsewhere than in the lockers provided, (2) If such accommodation is in a food room it shall be in the form of lockers or cupboards.

Regulation 21 Facilities for Washing food and equipment. (1) Sinks or other washing facilities suitable and sufficient for any necessary washing of food and equipment used in the food business, shall be provided in all food premises where open foods is handled, (2) Every such sink shall be provided with an adequate supply of hot and cold water or 'blended' water; or cold water only where the sink is used — (a) only for washing fish, fruit or vegetables; or (b) for washing with a suitable bactericidal agent only drinking vessels or only ice cream formers or servers, (3) Every such sink shall be kept clean and in good working condition.

Regulation 22 Lighting of food rooms. Suitable and sufficient means of lighting shall be provided in every food room and every such room shall be suitably and sufficiently lighted.

Regulation 23 Ventilation of food rooms. Except in the case of a room in which the humidity or temperature is controlled, suitable and sufficient means of ventilation shall be provided and maintained in every food room.

Regulation 25 Cleanliness and Repair of food rooms. Walls, floors, doors, windows, ceiling, woodwork and all other parts of the structure of every food room shall be kept clean, in good order, repair and condition — to prevent infestation.

Regulation 26 Accumulation of Refuse etc. (1) Layout of food premises to provide adequate space for waste removal and separation of unfit food and storage of same, (2) Solid or liquid refuse or filth to be not allowed to accumulate in food rooms etc.

PRIVATE DWELLINGS

Recommendations of BS.6465 Part 1: 1984 "Sanitary Installations" states that every new dwelling should be provided with a minimum of one WC, one bath or shower, one wash basin and one sink. Where a WC is separate, a wash basin should be provided in the same compartment, or in a room giving direct access to the WC compartment. A second WC with wash basin is a desirable provision with its location near the entrance to the dwelling. A bathroom containing a WC should not be entered directly from a bedroom unless it is intended for the sole use of the bedroom occupants, and a second WC is provided elsewhere in the dwelling.

Scale of provision of sanitary appliances

Type of dwelling	Appliances	Dwelling suitable for up to:				
		2 persons	3 persons	4 persons	5 persons	6 + persons
On one level e.g. bungalows or flats	WC	1*	1*	1 +	1 +	2‡
	Bath	1	1	1	1	1
	Wash basin	1	1	1	1	1
		And, in addition, one in every separate WC compartment which does not adjoin a bathroom				
	Sink and drainer	1	1	1	1	1
On two or more levels e.g. houses or maisonettes	WC	1*	1*	1 +	2‡	2‡
	Bath	1	1	1	1	1
	Wash basin	1	1	1	1	1
		And, in addition, one in every separate WC compartment which does not adjoin a bathroom				
	Sink and drainer	1	1	1	1	1

* May be in bathroom
+ Recommended to be in a separate compartment
‡ Of which, one may be in a bathroom

Note: In the private sector, a shower is equal to a bath (Public Health Act 1961, Section 33). In Scotland, a wash basin is required in, or adjacent to, every WC compartment (The Building Standards (Scotland) Regulations 1981/Q7).

In Scotland, a shower may be provided instead of a bath in all dwellings — both private and public sectors (The Building Standards (Scotland) Regulations 1981/Q7[1]).

The table is based on Circular 36/37 of the Ministry of Housing and Local Government and new Scottish Housing Handbook Bulletin No. 1, "Metric Space Standards".

Tables also provided in BS.6465: Part 1

— Accommodation for elderly people
— Residential homes for the elderly
— Office buildings for shops
— Factories
— Schools and higher educational establishments
— Cinemas, concert halls, theatres and similar buildings used for public entertainment
— Hotels
— Restaurants and canteens
— New public houses
— Swimming pools

Sanitation: Numerical Provision

FACTORIES

Statutory provision

Authority
The Sanitary Accommodation Regulations SRO 1938 No. 611 Factories, made under section 7(2) of the Factories Act 1937 by the Secretary of State. (Continues in force by virtue of Sixth Schedule para. 2 Factories Act 1961).

Females (employed). Minimum of one sanitary convenience for every twenty-five females (other numbers less than 25 to be reckoned as 25). This provision remains at 1 per 25 irrespective of the number of females employed.

Males (employed). As for females (given above), but conveniences must not be conveniences suitable merely as urinals. The provision of urinals is optional and additional to any conveniences provided. If number of males employed exceeds one hundred it is permissible to have one sanitary convenience for every 25 males up to the first hundred males; then one for every forty males thereafter providing sufficient urinal accommodation is also provided (numbers less than 40, to be reckoned as 40).

Further provisions for factories but applicable only to factories constructed, enlarged or converted for use as a factory before 1st July 1938 — where the number of males employed exceeded five hundred. Under these circumstances, providing there was sufficient urinal accommodation, it was sufficient to allow for one sanitary convenience for every sixty employed males. There are other conditions attached.

Interpretation
For the purpose of the Factories Act 1961, the expression 'sanitary convenience' is interpreted as including urinals, water closets, earth closets, privies, ash pits, and any similar convenience (Section 176 p.115)

Example of calculation for a new factory employing 210 females and 490 males (minimum requirements)

Females — one per 25, therefore 9 sanitary conveniences are required (210÷25=8 plus 1 for the odd 10=9).

Males — one per 25 for the first 100 = 4 sanitary conveniences. One per 40 for remainder = 10 sanitary conveniences (390÷40=9 plus 1 for the odd 30=10). Total for males = 14 sanitary conveniences plus sufficient urinal accommodation.

Note: No legal guidance is given as to what is regarded as 'sufficient' urinals but see CP3 (below) or Shops, Offices etc. Sanitary Accommodation Regulations 1964 for suggested provisions (e.g. 4 for the first 100 men, 3 for second 100, and 2½ for all the remaining 100's). Using either as a guide, 14 or 15 urinal stalls are indicated in the above example. For interpretation of expression 'Factory', see Factories Act 1961 Section 175.

AUTHORITATIVE PROVISION

Authority BS.6465: Part 1: 1984 Code of Practice for scale of provision selection and installation of sanitary appliances.

Fitment	For male personnel	For female personnel
Water Closets Note: A cleaners' sink should be fitted adjacent to sanitary accommodation (CP305)	1 for 1-15 persons 2 for 16-35 persons 3 for 36-65 persons 4 for 66-100 persons From 101-200 add at rate of 3% For over 200 add at rate of 2½%	1 for 1-12 persons 2 for 13-25 persons 3 for 26-40 persons 4 for 41-57 persons 5 for 58-77 persons 6 for 78-100 persons From 101-200 add at rate of 5% For over 200 add at rate of 4%
Urinals	Nil up to 6 persons 1 for 7-20 persons 2 for 21-45 persons 3 for 46-70 persons 4 for 71-100 persons From 101-200 add at rate of 3% For over 200 add at rate of 2½%	Nil
Washbasins or Ablution Fountains	1 per 15 up to 105 persons. For over 105, add at the rate of 5 per cent 1 per 20 persons	1 per 15 up to 105 persons. For over 105, add at the rate of 5 per cent 1 per 20 persons
Baths but preferably showers	As required for particular trades or occupations	

Example of calculation using same problem as dealt with above

Females (210)	W.C.'s	12	(6 for first 100, 5 for second 100, plus 1 for remainder)
	Wash basins	13	(7 for first 105, 5 for second 100, plus 1 for remainder)
Males (490)	W.C.'s	16	(4 for first 100, then 3 for each of next 3 ×100's, plus 3 for remainder)
	Wash basins		As for females
	Urinals	14	(4 for first 100, 3 for second 100, 2½ for each of next 2×100's, plus 2* for the remainder.

Before using the above table, attention should be drawn to the statutory requirements of the Factories Act 1961 and regulations made thereunder; and for the many trades of a dirty or dangerous character requiring more extensive provision of washing and bathing accommodation by law.

*Could be 3, making a total of 15.

Offices, Shops and Railway Premises

STATUTORY PROVISION

Authority
The Washing Facilities Regulations 1964 Shops and Offices S.I. 1964 No. 965 made under Sections 10 and 80 (3) of the Offices, Shops and Railway Premises Act 1963 and The Sanitary Conveniences Regulations 1964 Shops and Offices S.I. 1964 No. 966 made under Sections 9 and 80 (3) of the same Act, by the Minister of Labour.

SEPARATE ACCOMMODATION FOR PERSONS OF EACH SEX

Washing facilities
Regulation No. 3 and Schedule Parts I and II
Washing facilities may be shared providing the number of persons regularly employed to work therein* does not exceed five. Over five persons, proper separate accommodation shall be provided unless the circumstances affecting the premises are such that it is not reasonably practicable to provide washing facilities in proper separate accommodation for each sex.

Sanitary conveniences
Regulation No. 3 and Schedule Part 1

Water closets and chemical closets may be shared providing the number of persons regularly employed to work therein* does not exceed five, or each of the regular employees normally works in the premises for only 2 hours or less daily — only one WC and one basin need be provided for use by all staff.

Over five persons, accommodation must be separate.

Note: There is no question of reasonability as for washing facilities.

*The term "therein" is meant to imply the premises.

REQUIREMENTS FOR PREMISES NOT EXEMPTED

A Working Hours or number of Employees	Washing facilities	Sanitary Conveniences
Where daily working hours of any employee does not exceed two or the number of employed persons does not exceed five	One wash basin or trough or washing fountain	One water closet
(Whether or not persons of both sexes are employed)	See Schedule Pt 1 Para 1	See Schedule Pt 1 Para.1

B Premises other than those in 'A	Male and	Female
Number of regularly employed persons	Wash basins or + Units of Trough or + Units of Washing fountain	Water Closets Note For males, use this Table when urinals are not provided to the scale in Table C
1 to 15 16 to 30 31 to 50 51 to 75	1 Separate 2 accommodation, 3 may be combined 4 (see Regulation 3 above)	1 Separate male 2 and female 3 accommodation 4 shall be
76 to 100 Exceeding 100 + One unit = 0.6 m of length or circumference	5 Add at the rate of 1 per 25 persons Note: Fractions of 25 to be treated as 25 in all cases See Schedule Pt. 1 para 2	5 provided See Schedule Pt 1, Para 2a and b

C Premises other than those in 'A'	Males only Note: Use this Table when urinal accommodation is provided	
Number of regularly employed males	Water Closets	Units of Urinal accommodation
1 to 15	1	—
16 to 20	1	1
21 to 30	2	1
31 to 45	2	2
46 to 60	3	2
61 to 75	3	3
76 to 90	4	3
91 to 100	4	4
Exceeding 100 (See Schedule Pt 1 Para.2c)	Add at the rate of 1 WC or urinal per 25 males but not less than ¾ of the additional sanitary conveniences shall be WC's	

PREMISES EXEMPTED FROM CERTAIN REQUIREMENTS

Exemption for which the premises may qualify. See O.S.R. 1963 (Exemption No. 1 Order) 1964	Fitment	Situation A	Situation B	Situation C
		Where the daily working hours of any employee does not exceed two	Where the number of employed persons does not exceed five	Subject to Situation A, plus the number of employed persons exceeding five
		Number of fitments	Number of fitments	Number of fitments
Premises exempted from the requirement to supply running water	Fixed or portable washbowl	One Whether or not persons of both sexes are employed (Washing facilities Regs. Schedule Pt. II)	One Whether or not persons of both sexes are employed	One per 'unit' of five persons Where separate accommodation is provided for each sex, the 'units' refer to persons of each sex
Premises where it is not reasonably practicable to provide a drainage system and a supply of water for flushing purposes	Chemical closet	One Whether or not persons of both sexes are employed (Sanitary Conveniences Regs. Schedule Pt. II)	One Whether or not persons of both sexes are employed	Number of persons of each sex — Closets 1 to 15 persons — 1 16 to 30 persons — 2 31 to 50 persons — 3 51 to 75 persons — 4 76 to 100 persons — 5 Note: Accommodation must be separate for the sexes

The term "wash bowl" is interpreted in the Regulations as 'including any water container suitable for use as a washing facility.'

Sanitation: Numerical Provision

FURTHER STATUTORY PROVISIONS

Authority
The Washing Facilities Regulations 1964 and Sanitary Conveniences Regulations 1964

Points concerning excluded premises
Although the Offices, Shops and Railway Premises Act 1963 has extensive coverage, certain kinds of premises are not included. The following are some examples. (i) Premises where only self-employed people work; (ii) Businesses where the only people employed are any of the following immediate relatives of the employer — husband, wife, parent, grandparent, son, daughter, grandchild, brother or sister; (iii) Premises where the sum of hours worked by all employees is normally not more than 21 hours per week e.g. if a self-employed person employs only a part-time cleaner or an assistant on one or two days per week, the premises are not covered by the Act unless the total of all hours worked by the employees normally exceeds 21 each week.

SCALE OF PROVISION OF SANITARY APPLIANCES FOR OFFICE BUILDINGS AND SHOPS

Authority BS6465: Part 1: 1984 Code of Practice for scale of provision, selection and installation of sanitary appliances.

Appliances	Accommodation other than principals, etc.	
	For male staff	For female staff
WC's* (No urinals provided)	1 for 1 — 15 persons 2 for 16 — 30 persons 3 for 31 — 50 persons 4 for 51 — 75 persons 5 for 76 — 100 persons For over 100 persons, add one unit for every additional 25 persons or part thereof	1 for 1 — 15 persons 2 for 16 — 30 persons 3 for 31 — 50 persons 4 for 51 — 75 persons 5 for 76 — 100 persons For over 100 persons, add one unit for every additional 25 persons or part thereof
WC's* (Urinals provided)	1 for 1 — 20 persons 2 for 21 — 45 persons 3 for 46 — 75 persons 4 for 76 — 100 persons For over 100 persons, add one unit for every additional 24 persons or part thereof, but one in four of the additional fitments may be a urinal	
Urinals* (1 stall or 600 mm of space)	0 for 1 — 15 persons 1 for 16 — 30 persons 2 for 31 — 60 persons 3 for 61 — 90 persons 4 for 91 — 100 persons For over 100 persons, additional provision is determined by the number of WC's. (See previous item)	
Wash basins, trough or washing fountains **	1 for 1 — 15 persons 2 for 16 - 30 persons 3 for 31 — 50 persons 4 for 51 — 75 persons 5 for 76 — 100 persons, For over 100 persons add one unit for every additional 25 persons or part thereof.	Provision as for males
Bins, incinerators or macerator units for sanitary dressings disposal		1 in every sanitary accommodation regularly used by females
Cleaner's sink	At least 1 per floor, preferably in or adjacent to a sanitary apartment	

*Sanitary Convenience Regulations 1964 (St. Inst. No. 966)
** Washing Facilities Regulations 1964 (St. Inst. No. 965)
Note: The scale of provision shown in the table should be increased by one (and in male toilets a single bowl urinal should be provided) if the sanitary conveniences serve a staff of more than 10 persons and are also used by members of the public visiting the premises.

The table is based on the requirements of the Offices, Shops and Railway Premises Act 1963.

Before using the results obtained from this table, they should be checked for compliance with the minimum requirements of the two regulations quoted above.

SHARING FACILITIES

Arrangements may be made for all or any of the persons employed to work in the premises to have the use of washing facilities provided primarily for the use of others. In determining the number of fitments (wash-basins, wash-bowls, units of trough or washing fountain, or sanitary conveniences) the combined total number of persons for whose regular use the said facilities are made available is the relevant number (see Regulation 4 in both Regulations and Section 10(5) in the main Act 1963).

FACILITIES USED BY THE PUBLIC

In the case of premises in which the number of persons employed to work therein at any one time regularly exceeds ten, and washing facilities and sanitary conveniences (provided primarily for the use of employees) are also made available for use by members of the public resorting to the premises, the total number of wash basins etc. and water closets etc., shall in every case be increased by one (see Regulation 5 in both Regulations referred to.)

DISPOSAL OF SANITARY DRESSINGS

Where the total number of females (not being members of the general public) for whose regular use sanitary conveniences are made available exceed ten, suitable and effective means of disposal of sanitary dressings shall be provided. All means so provided shall be constantly maintained in proper condition. Where the means provided consist or include bins, the contents of such bins shall be disposed of at suitable intervals (see Regulation 9 in Sanitary Conveniences Regulations).

Schools and Refreshment Houses

STATUTORY PROVISION FOR SCHOOLS

Authority
The Education (School Premises)
Regulations 1981

Note: These Regulations do not lend themselves to easy tabulation but the following table is compiled from the Regulations and should serve as a useful guide. In all cases the Regulations should be referred to if any point requires clarification or amplification.

Scale of provision of sanitary appliances

Table 6 from BS.6465 — minimum requirements:

Schools and higher educational establishments

Appliances	Special Schools	Primary schools	Secondary schools
All 'fittings' i.e. WCs + urinals	1/10 of the number of pupils Rounded up to nearest whole even number	Aggregate of 1/10 of the number of pupils under 5 and 1/20 of the number of others. Not less than 4. Rounded up to nearest whole even number	1/20 of the number of pupils. Not less than 4. Rounded up to nearest whole even number
WCs only	Girls: all fittings	Girls: all fittings	Girls: all fittings
Urinals	Boys: not more than 2/3 of fittings may be urinals	Boys: not more than 2/3 of fittings may be urinals	Boys: not more than 2/3 of fittings may be urinals
Wash basins	As primary or secondary schools	Not less than the number of fittings	1 in each washroom, at least 2 basins per 3 sanitary fittings
Showers	*	*	Sufficient for physical education
Wash basin (medical inspection)	At least 1, accessible to WC	At least 1, accessible to WC	At least 1, accessible to WC
Cleaners'/slop sink	At least one per floor	At least one per floor	At least one per floor

*Whilst not required by statute, it is suggested that sufficient showers should be provided for physical education.

Nursery schools and play schools

WCs	1 per 5 boarding pupils
Wash basins	1 per 3 pupils for first 60 boarding pupils 1 per 4 pupils for next 40 boarding pupils 1 per 5 pupils for every additional 5 boarding pupils
Baths	1 per 10 boarding pupils
Showers	May be provided as alternative to not more than 3/4 of the minimum number of baths

Boarding schools†

WCs	1 per 10 pupils (not less than 4)
Wash basins	1 per WC
Sinks	1 per 40 pupils

† Where sanitary accommodation for day pupils is accessible to, and suitable for the needs of, boarders, these requirements may be reduced to such an extent as may be approved in each case.

Sanitation: Numerical Provision

AUTHORITATIVE PROVISION FOR REFRESHMENT HOUSES

Notes: Some teaching and other staff may be subject to Regulations under the Offices, Shops and Railway Premises Act, but for all members the Regulations should be used as a guide; see table 4.

Attention is drawn to the necessity to provide facilities for the disposal of sanitary dressings.

For educational establishments not mentioned above reference should be made to the Department of Education and Science.

Requirements in Scotland are based upon the School Premises (General Requirements and Standards) (Scotland) Regulations 1967 and 1973.

Requirements in Northern Ireland are based upon the statutory requirements of the Department of Education (Northern Ireland).

This table is based on the Education (School Premises) Regulations 1981 (S.I.909).

A general requirement common to all schools is that there shall be separate, sufficient and suitable sanitary washing accommodation for teaching staff and school meals service staff. Medical inspection rooms should have a wash basin and be conveniently accessible to a closet. Also not included in the Regulations, the provision of 1 cleaners' sink per floor is recommended.

Authority
Public Health Act 1936 Section 89 'Power to require sanitary conveniences to be provided at inns, refreshment houses, etc.' (also includes 'place of public entertainment'). Public Health Act 1961 Section 80 gives meaning of "refreshment house" in Section 89 as 'any building in which food or drink is sold to and consumed by the public'.

Note: No statutory numerical provision applicable throughout the entire country is in existence.

MODEL STANDARD PROVISION FOR CARAVAN SITES

Authority

Model Standards specified by the Minister of Housing and Local Government in pursuance of his power under Section 5(6) of the Caravan Sites and Control of Development Act 1960. They represent the standards normally to be expected, as a matter of good practice, on sites which are used regularly by residential or holiday caravans. They should be applied with due regard to the particular circumstances of each case.

Permanent residential caravan sites

Drainage, sanitation and washing facilities.

Para. 8. Satisfactory provision should be made for foul drainage, either by connection to a public sewer or by discharge to a properly constructed septic tank or cesspool.

Para. 9. For caravans having their own water supply and WC's, each caravan standing should be provided with a connection to the foul drainage system, the connection should be capable of being made air tight when not in use. For caravans without such facilities, communal toilet blocks should be provided, with adequate supplies of water, on at least the following scales:

Men: 1 WC and 1 Urinal per 15 caravans
Women: 2 WC's per 15 caravans
1 wash basin for each WC or group of WC's
1 shower or bath (with hot and cold water) for each sex per 20 caravans

Para. 10. Laundry facilities should be provided, in a separate room, minimum 1 deep sink per 15 caravans (with running hot and cold water).

Para. 11. Properly designed disposal points for the contents of chemical closets should be provided, with an adequate supply of water for cleaning the containers.

Holiday caravan sites

The foregoing standards should apply, subject to the following modifications:

2. Water supply and drainage connections to individual caravan standings etc., may be dispensed with.

4. Wash basins should be provided on a scale of not less than one for men and one for women per 15 caravans.

5. Laundry facilities should be provided on a scale of minimum 1 deep sink with running hot and cold water per 30 caravans.

PROVISION FOR SPORTS PAVILIONS

Females 1 WC for each ten women.
Males 2 WC's and 4 urinal stalls for the first 25 men, then 1 WC and 2 urinal stalls for every additional 25 men.

Places of Entertainment

PROVISION FOR BUILDINGS USED FOR PUBLIC ENTERTAINMENT

Appliance	For male public	For female public
WC's in theatres, concert halls and similar premises	Minimum of 1 for up to 250 males plus 1 for every additional 500 males or part thereof	Minimum of 2 for up to 50 females. 3 for 51-100 females plus 1 for every additional 40 females or part thereof
WC's in cinemas	Minimum of 1 for up to 250 males, plus 1 for every additional 500 males or part thereof	Minimum of 2 for up to 75 females. 3 for 76-150 females plus 1 for every additional 80 females or part thereof
Urinals in theatres, concert halls and similar	Minimum of 2 for up to 100 males, plus 1 for every additional 80 males or part thereof	
Urinals in cinemas	Minimum of 2 for up to 200 males, plus 1 for every additional 100 males or part thereof	
Wash basins	1 per WC and, in addition, 1 per 5 urinals or part thereof	1 plus 1 per 2 WC's
Cleaner's sinks	Adequate provision should be made for cleaning facilities, including at least 1 cleaner's sink	

Where reliable estimates are not available, assume audience consisting of 50 per cent male and 50 per cent female.

Note: Facilities should be provided for disposal of sanitary dressings. Staff may be subject to Offices, Shops and Railways Premises Act. Sanitary accommodation for performers should be based on Offices, Shops and Railway Premises Act.

Hotels, Restaurants and

AUTHORITATIVE PROVISION FOR HOTELS

Authority BS.6465:1984 Table 8, and the
Offices, Shops and Railways Act

Fitments	For residential public and staff	For public rooms*		For non-residential staff	
		For Males	For Females	For male staff	For female staff
Water closets	1 per 9 persons omitting occupants of rooms with WCs 'en suite'	1 per 100 up to 400. For over 400 add at the rate of 1 per 250 or part thereof	2 per 100 up to 200. For over 200 add at the rate of 1 per 100 or part thereof	1 for 1 — 15 persons 2 for 16 — 35 persons 3 for 36 — 65 persons 4 for 66 — 100 persons	1 for 1 — 12 persons 2 for 13 — 25 persons 3 for 26 — 40 persons 4 for 41 — 57 persons 5 for 58 — 77 persons 6 for 78 — 100 persons
Urinals		1 per 50 persons		Nil up to 6 persons 1 for 7 — 20 persons 2 for 21 — 45 persons 3 for 46 — 70 persons 4 for 71 — 100 persons	
Wash basins	1 per bedroom and at least 1 per bathroom			1 for 1 — 15 persons 2 for 16 — 35 persons 3 for 36 — 65 persons 4 for 66 — 100 persons	1 for 1 — 12 persons 2 for 13 — 25 persons 3 for 26 — 40 persons 4 for 41 — 57 persons 5 for 58 — 77 persons 6 for 78 — 100 persons
Bathrooms	1 per 9 persons omitting occupants of rooms with baths 'en suite'	*Note: When assessing numerical provision for public rooms it may be assumed that there will be equal numbers of males and females.			
Cleaners sinks	1 per 30 bedrooms minimum 1 per floor				

Sanitation: Numerical Provision

AUTHORITATIVE PROVISION FOR RESTAURANTS AND CANTEENS

Authority BS.6465:1984 Table 9, and
the Offices, Shops and Railways Act

Fitments	*For male public	*For female public	For male staff	For female staff
Water closets	1 per 100 up to 400. For over 400, add at the rate of 1 per 250 or part thereof	2 per 100 up to 200. For over 200, add at the rate of 1 per 100 or part thereof	1 for 1 — 15 persons 2 for 16 — 35 persons 3 for 36 — 65 persons 4 for 66 — 100 persons	1 for 1 — 12 persons 2 for 13 — 25 persons 3 for 26 — 40 persons 4 for 41 — 57 persons 5 for 58 — 77 persons 6 for 78 — 100 persons
Urinals	1 per 25 persons		Nil up to 6 persons 1 for 7 — 20 persons 2 for 21 — 45 persons 3 for 46 — 70 persons 4 for 71 — 100 persons	
Wash basins	1 per WC and 1 per 5 urinals	1 per 2 WC's	1 for 1 — 15 persons 2 for 16 — 35 persons 3 for 36 — 65 persons 4 for 66 — 100 persons	1 for 1 — 12 persons 2 for 13 — 25 persons 3 for 26 — 40 persons 4 for 41 — 57 persons 5 for 58 — 77 persons 6 for 78 — 100 persons

HOSPITALS

The requirements for sanitary appliances in hospitals and other health buildings are covered by the general recommendations of BS.6465. However, for the scale of provision, ergonomic data and the special requirements for appliances in hospitals — the various guidance documents produced by the Department of Health and Social Security should be followed. These include: Building Notes — especially 'common places'; Activity Data base; and Component Data base.

Note: Facilities for the disposal of sanitary dressings should be provided at the above premises

Symbols for Drawings & References

FITMENT SYMBOLS FOR SMALL SCALE DRAWINGS

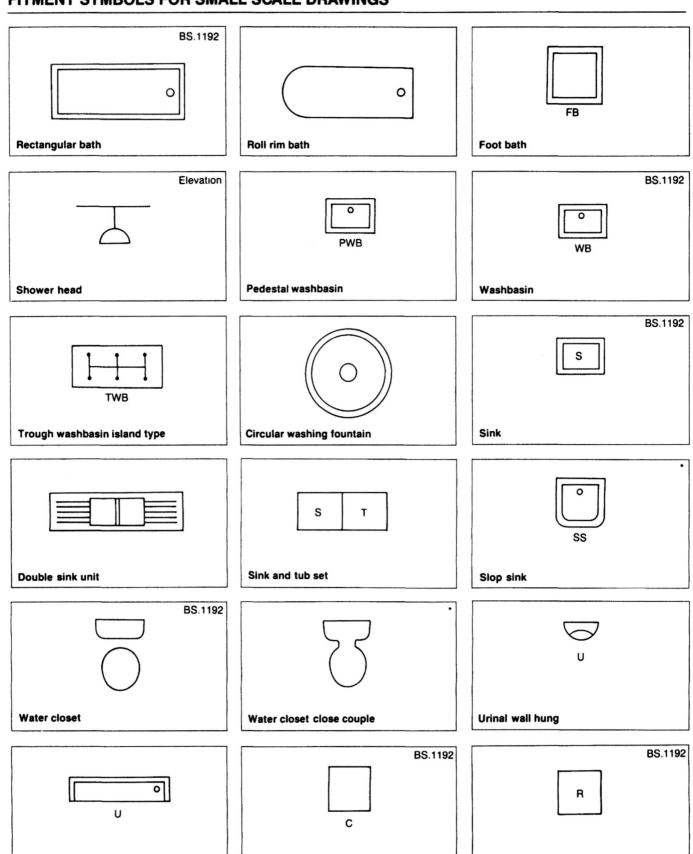

BS.1192

Rectangular bath

Roll rim bath

Foot bath — FB

Shower head — Elevation

Pedestal washbasin — PWB

Washbasin — WB — BS.1192

Trough washbasin island type — TWB

Circular washing fountain

Sink — S — BS.1192

Double sink unit

Sink and tub set — S | T

Slop sink — SS

Water closet — BS.1192

Water closet close couple

Urinal wall hung — U

Slab urinal — U

Cooker — C — BS.1192

Refrigerator — R — BS.1192

Sanitation: Layout Planning

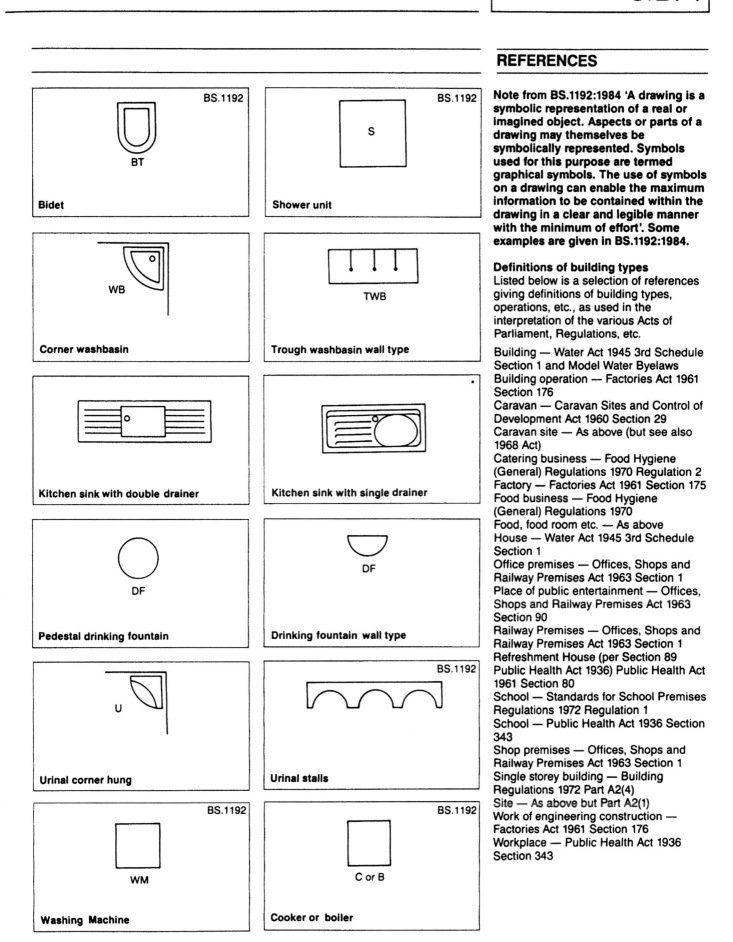

Note from BS.1192:1984 'A drawing is a symbolic representation of a real or imagined object. Aspects or parts of a drawing may themselves be symbolically represented. Symbols used for this purpose are termed graphical symbols. The use of symbols on a drawing can enable the maximum information to be contained within the drawing in a clear and legible manner with the minimum of effort'. Some examples are given in BS.1192:1984.

Definitions of building types
Listed below is a selection of references giving definitions of building types, operations, etc., as used in the interpretation of the various Acts of Parliament, Regulations, etc.

Building — Water Act 1945 3rd Schedule Section 1 and Model Water Byelaws
Building operation — Factories Act 1961 Section 176
Caravan — Caravan Sites and Control of Development Act 1960 Section 29
Caravan site — As above (but see also 1968 Act)
Catering business — Food Hygiene (General) Regulations 1970 Regulation 2
Factory — Factories Act 1961 Section 175
Food business — Food Hygiene (General) Regulations 1970
Food, food room etc. — As above
House — Water Act 1945 3rd Schedule Section 1
Office premises — Offices, Shops and Railway Premises Act 1963 Section 1
Place of public entertainment — Offices, Shops and Railway Premises Act 1963 Section 90
Railway Premises — Offices, Shops and Railway Premises Act 1963 Section 1
Refreshment House (per Section 89 Public Health Act 1936) Public Health Act 1961 Section 80
School — Standards for School Premises Regulations 1972 Regulation 1
School — Public Health Act 1936 Section 343
Shop premises — Offices, Shops and Railway Premises Act 1963 Section 1
Single storey building — Building Regulations 1972 Part A2(4)
Site — As above but Part A2(1)
Work of engineering construction — Factories Act 1961 Section 176
Workplace — Public Health Act 1936 Section 343

APPLIANCES AND SPACE ALLOWANCES

Bath

'n' — any natural number

Co-ord. size 700mm (+ 'n' × 100)

Co-ord. size 1700mm (+n × 100)

Height: Co-ordinating size = 'n' × 50 typical bath has adjustable height between 440mm and 515mm. Longer 'feet bolts' available to give extra height for fitting 75mm seal trap above the floor

Water closets

520mm

745mm

Heights various

Low level

520mm

650mm

Measurements given for W.C.'s are examples only

356 mm

High level

Wash basin

B

A

A 560mm — 635mm

B 405mm — 460mm

Height to top of front edge — 785mm

Bidet

545 mm

Height — 380mm

Sinks

455 mm

'Belfast' or 'London' type

610mm

1000mm

Single Drainer

600 or 500mm

1500mm

Double drainer

Height: 900mm

PLANNING BY EXAMPLE

Examples of planning for single stack drainage

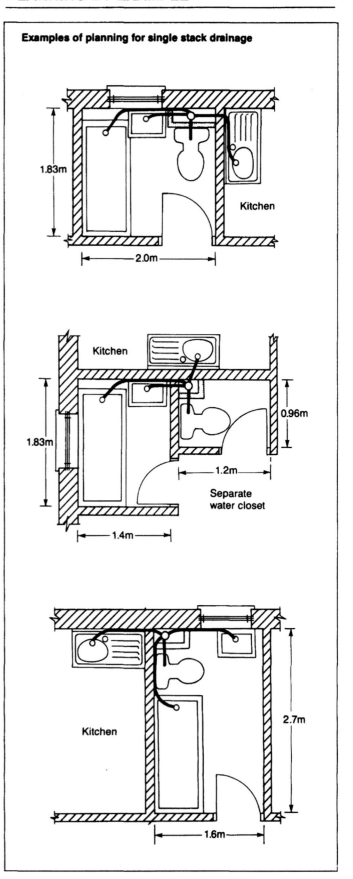

1.83m

2.0m

Kitchen

Kitchen

1.83m

0.96m

1.2m

1.4m

Separate water closet

2.7m

Kitchen

1.6m

EXAMPLE OF DOUBLE GROUP PLANNING

FLATLET BATHROOMS

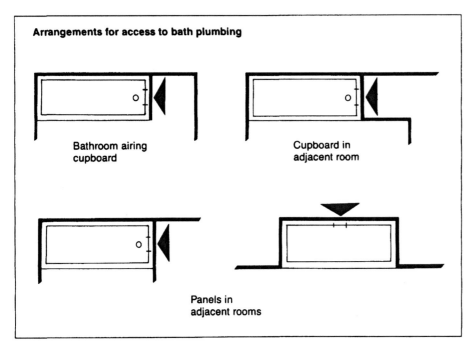

Examples of double group planning for single track drainage

Kitchen

1.83m

1.83m

Kitchen

Showing alternative positions for the sink unit

Duct

Access panel

0.46m

Central corridor

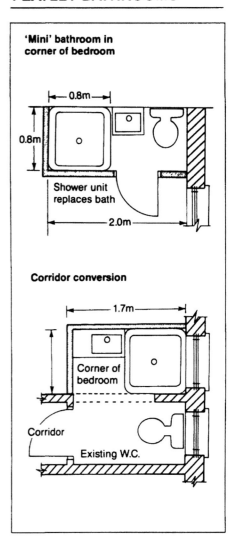

'Mini' bathroom in corner of bedroom

0.8m

0.8m

Shower unit replaces bath

2.0m

Corridor conversion

1.7m

Corner of bedroom

Corridor

Existing W.C.

Arrangements for access to bath plumbing

Bathroom airing cupboard

Cupboard in adjacent room

Panels in adjacent rooms

HOTELS

The following examples are given by Twyfords Bathrooms

Hotel bathroom

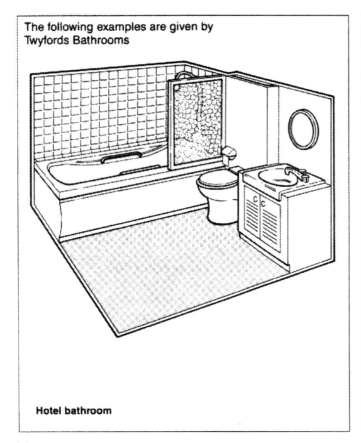

Hotel bathroom

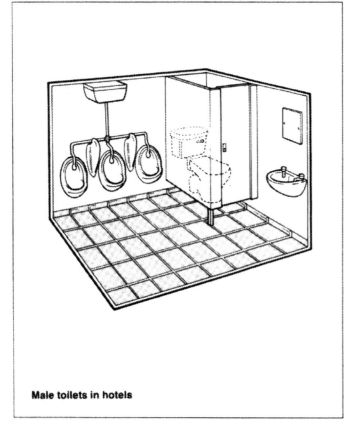

Male toilets in hotels

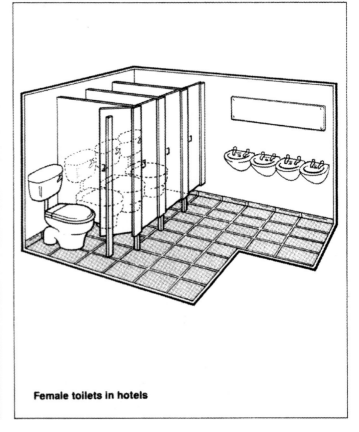

Female toilets in hotels

Sanitation: Layout Planning

RESTAURANTS AND CAFES

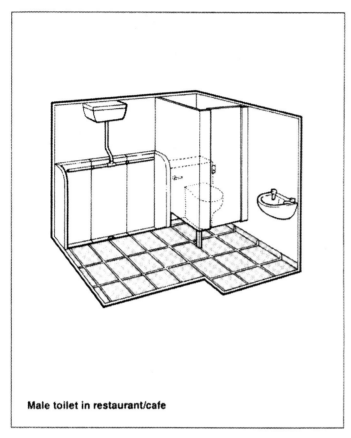

Male toilet in restaurant/cafe

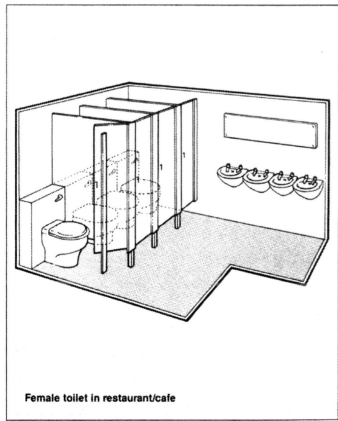

Female toilet in restaurant/cafe

OFFICES AND SHOPS

The following examples are given by
Twyfords Bathrooms

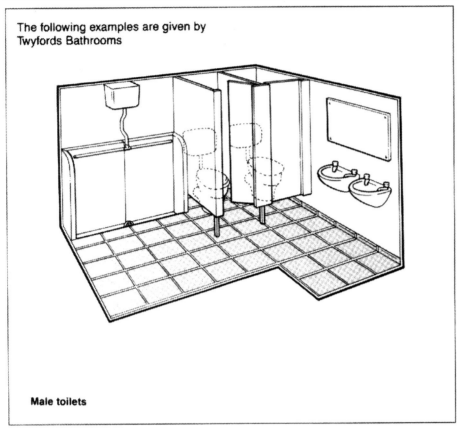

Male toilets

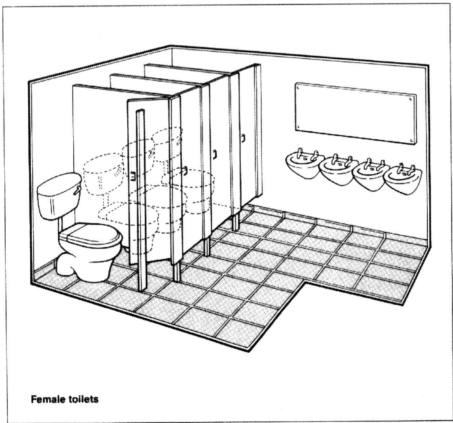

Female toilets

PUBLIC CONVENIENCES

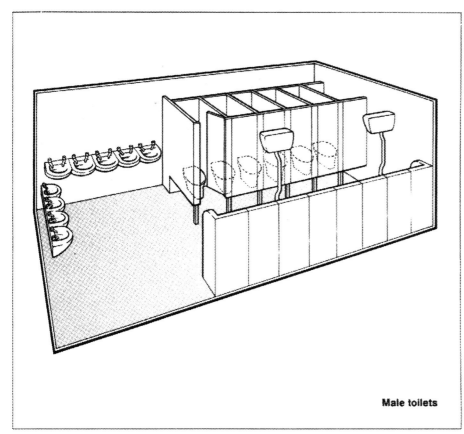

Male toilets

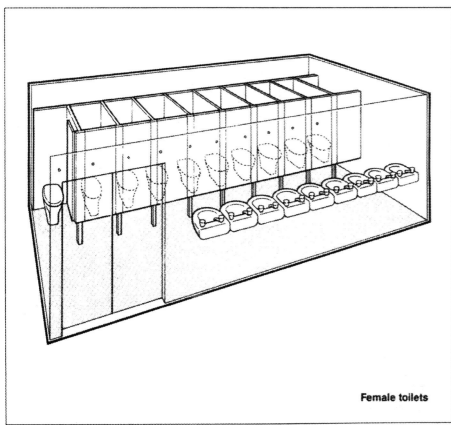

Female toilets

SCHOOLS

The following examples are given by
Twyfords Bathrooms

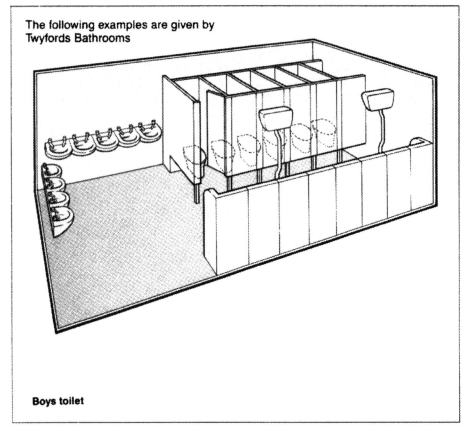

Boys toilet

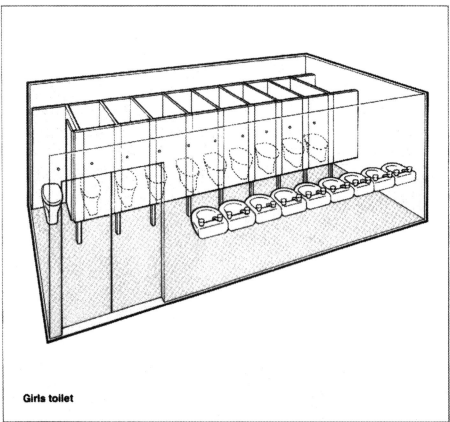

Girls toilet

Sanitation: Layout Planning

SPORTS CENTRES AND SWIMMING POOLS

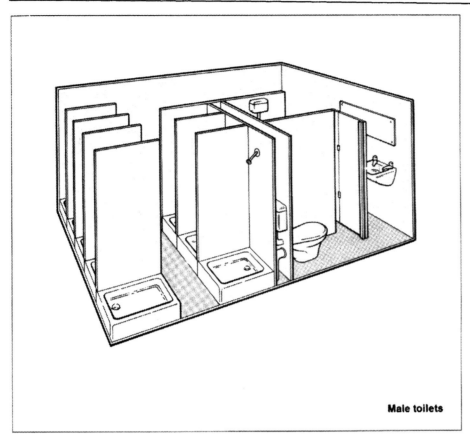

Male toilets

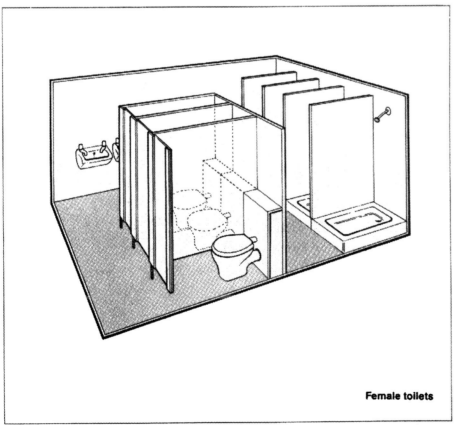

Female toilets

Facilities for Disabled People

The following information is taken from BS.5910:1979 "Access for the disabled to buildings". Clause 8 deals with lavatories.

Section 8.1 states that all lavatory accommodation accessible to wheelchair users should include an accessible W.C. compartment and allow for wheelchair approach to the other facilities such as hand basins, drinking fountains, incinerators, mirrors and towels.

The preferred W.C. compartment is illustrated right. It allows for the disabled person to be assisted by another person and is referred to as a "Unisex W.C. Facility". The dimensions should not be less than those indicated, and should be fitted with equipment as shown.

The code recommends the provision of at least one unisex W.C. facility in the following:

a) Public lavatory accommodation in shopping centres and large department stores; transport buildings such as air terminals, motorway service stations and principal railway stations; concert halls, etc.; as well as conference buildings, exhibition halls, sports and leisure centres, hospitals, health and welfare centres.

b) Staff lavatory accommodation in large office buildings, and other buildings where large numbers of people are employed.

Smaller W.C. accommodation can be provided where the size and layout in a building prevents a unisex W.C. Facility to be incorporated, or where a unisex W.C. facility to be incorporated, or where a unisex W.C. facility has been provided in one part of a building and further additional lavatory accommodation is required for wheelchair users.

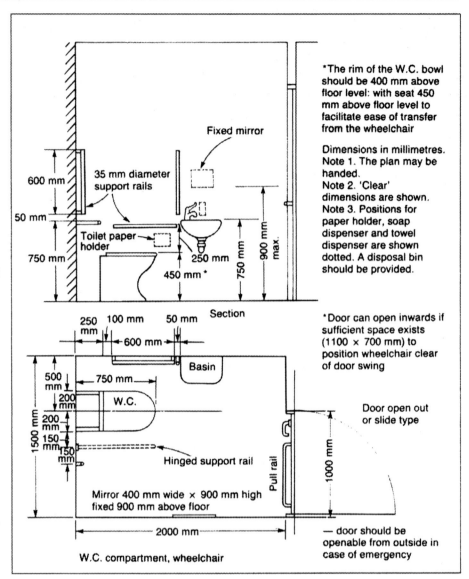

*The rim of the W.C. bowl should be 400 mm above floor level: with seat 450 mm above floor level to facilitate ease of transfer from the wheelchair

Dimensions in millimetres.
Note 1. The plan may be handed.
Note 2. 'Clear' dimensions are shown.
Note 3. Positions for paper holder, soap dispenser and towel dispenser are shown dotted. A disposal bin should be provided.

*Door can open inwards if sufficient space exists (1100 × 700 mm) to position wheelchair clear of door swing

Door open out or slide type

— door should be openable from outside in case of emergency

W.C. compartment, wheelchair

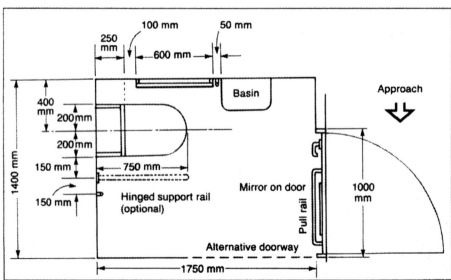

Sanitation: Layout Planning

Minimum space requirement for wheelchair W.C. compartment

Lobbies to lavatory accommodation
The following layouts should be provided where sanitary accommodation is accessible to wheelchair users through a lobby

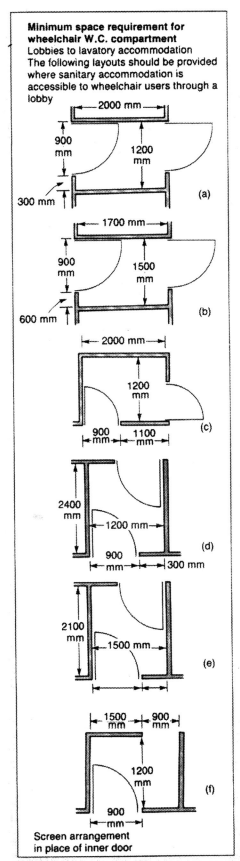

(a)

(b)

(c)

(d)

(e)

(f)

Screen arrangement in place of inner door

W.C. compartments for the ambulant disabled

These may be provided in large buildings where lavatory accommodation is provided at each floor but where a unisex W.C. facility is provided on one floor only: minimum overall dimensions

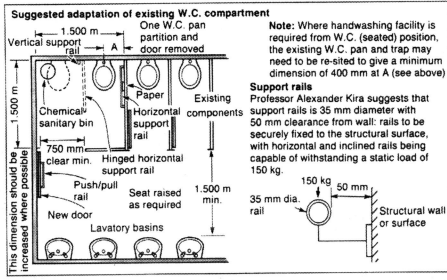

Section

35 mm diameter support rail

15°

250 mm

450 mm

400 mm approx.

Plan

800 mm

500 mm minimum

200 mm

1500 m

Door to open out or slide
be openable from outside in case of emergency

D.E.S. Design
Note 18 — Access for the physically disabled to educational buildings
— (7) Sanitary accommodation — provisions for access and lavatory facilities are similar to those in BS.5810:1979

Where adaptations of existing premises (schools) are undertaken conversions of the type illustrated are considered 'deemed to satisfy'

Suggested adaptation of existing W.C. compartment

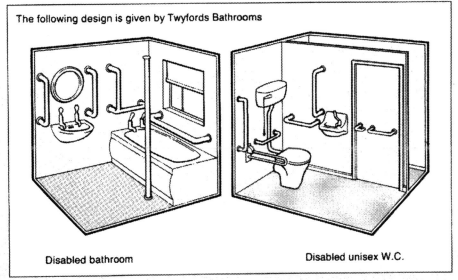

1.500 m
Vertical support rail
A
One W.C. pan partition and door removed
Paper
Horizontal support rail
Existing components

This dimension should be increased where possible
1.500 m
Chemical sanitary bin
750 mm clear min.
Hinged horizontal support rail
Push/pull rail
Seat raised as required
New door
Lavatory basins
1.500 m min.

Note: Where handwashing facility is required from W.C. (seated) position, the existing W.C. pan and trap may need to be re-sited to give a minimum dimension of 400 mm at A (see above)

Support rails
Professor Alexander Kira suggests that support rails is 35 mm diameter with 50 mm clearance from wall: rails to be securely fixed to the structural surface, with horizontal and inclined rails being capable of withstanding a static load of 150 kg.

150 kg
50 mm
35 mm dia. rail
Structural wall or surface

The following design is given by Twyfords Bathrooms

Disabled bathroom

Disabled unisex W.C.

Hospitals

The following examples are given by
Twyfords Bathrooms

Sanitation: Layout Planning

WARD TOILETS

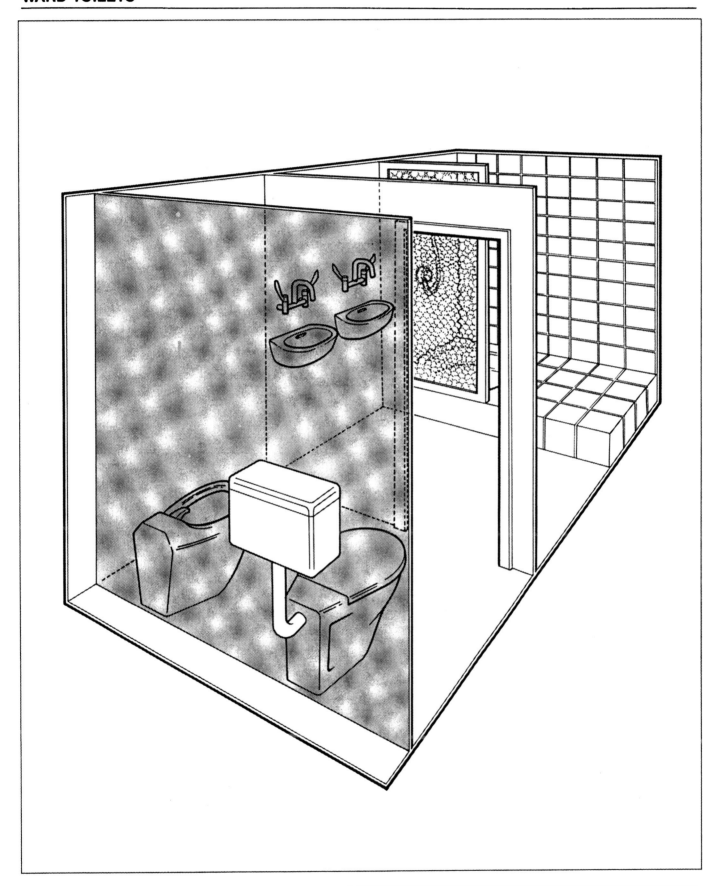

BUILDING REGULATIONS

Part F of Schedule I to the Building Regulations 1985 requires that there shall be means of ventilation so that an adequate supply of air may be provided for the people in the building. This is applicable to:
(a) dwellings
(b) buildings containing dwellings
(c) rooms containing sanitary conveniences
(d) bathrooms

ILLUSTRATING 'SANITARY ACCOMMODATION'

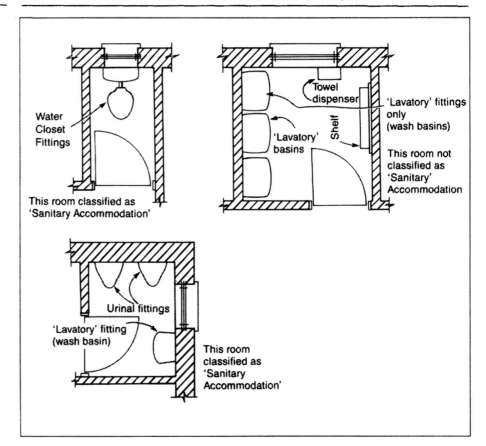

Approved Document F1 — Table 1 recommends that sanitary accommodation shall have a window, skylight or other similar means of ventilation which opens directly into the external air and of which the area capable of being opened is not less than one twentieth of the floor area; OR

Approved Document F1 — Table 2 states that mechanical means of ventilation which effects not less than three changes of air per hour and discharges directly into the external air.

Note: The ventilation may be intermittent but should run for at least 15 minutes after the use of the room or space stops.

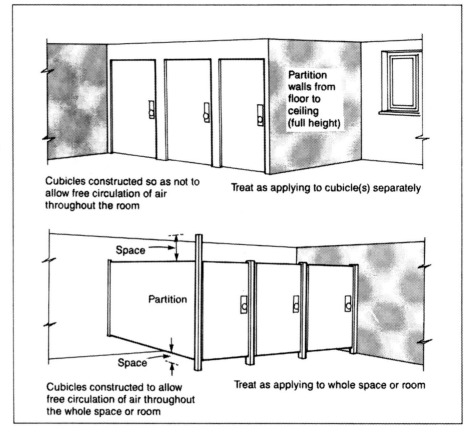

Sanitation: Ventilation of Sanitary Accom.

"Sanitary Accommodation" means a room or space constructed for use in connection with a building and which contains watercloset fittings or urinal fittings, whether or not it also contains other sanitary or lavatory fittings: Provided that if any such room or space contains a cubicle(s) so constructed as to allow free circulation of air throughout the room or space, then this regulation shall be treated as applying to the room or space as a whole and not to the cubicle or cubicles separately.

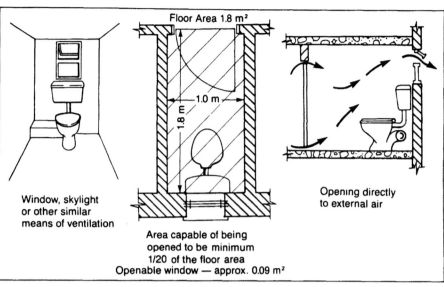

Window, skylight or other similar means of ventilation

Floor Area 1.8 m²

1.0 m

1.8 m

Area capable of being opened to be minimum 1/20 of the floor area
Openable window — approx. 0.09 m²

Opening directly to external air

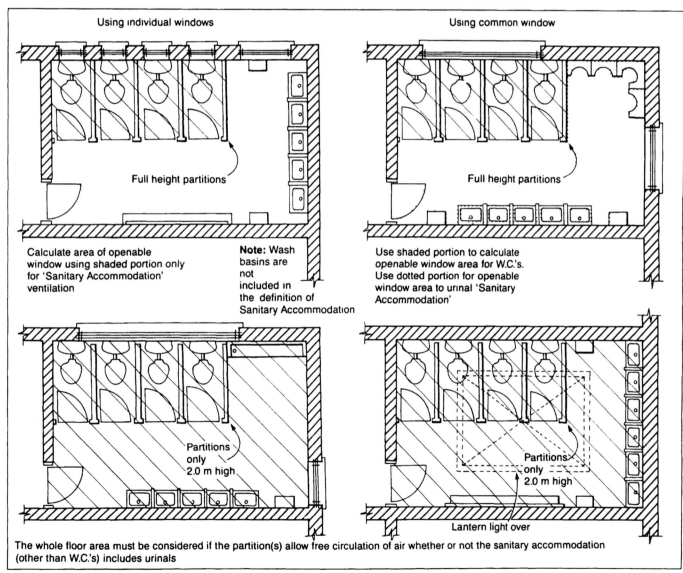

Using individual windows

Full height partitions

Calculate area of openable window using shaded portion only for 'Sanitary Accommodation' ventilation

Note: Wash basins are not included in the definition of Sanitary Accommodation

Partitions only 2.0 m high

Using common window

Full height partitions

Use shaded portion to calculate openable window area for W.C.'s. Use dotted portion for openable window area to urinal 'Sanitary Accommodation'

Partitions only 2.0 m high

Lantern light over

The whole floor area must be considered if the partition(s) allow free circulation of air whether or not the sanitary accommodation (other than W.C.'s) includes urinals

Statutory Requirements (2)

FACTORIES

vide Sanitary Accommodation Regulations 1938 Reg. 5 'Every sanitary convenience shall be sufficiently ventilated and shall not communicate with any workroom except through the open air or through an intervening ventilated space'. (Note: This regulation includes an exemption for workrooms in use prior to 1.1.1903 providing they are mechanically ventilated in a certain manner.)

The above regulation is applicable to all Factories as defined in Section 175 of the Factories Act, 1961.

OFFICES, SHOPS AND RAILWAY PREMISES

vide the Act of 1963
Section 9(2) 'Conveniences... shall be kept clean... and effective provision shall be made for lighting and ventilating them.

vide Sanitary Conveniences Regulations 1964
Reg. 6(2) Except as provided in para. 3 (below) — no sanitary accommodation (e.g. W.C.'s, urinals, etc.) which '... is not wholly enclosed shall be so situated that access to it is obtained directly from any room in which any person (other than a lavatory attendant) is employed to work.'

(3) The above requirement shall not apply where '(a) it is not reasonably practicable to comply' and (b) the sanitary accommodation 'was first installed or constructed before the date of making these Regs...' and (c) in any such case the 'accommodation shall be provided with effective mechanical means of ventilation which shall discharge directly into the open air and which shall be kept in operation during the periods during which any person is employed to work in the room from which access is obtained directly...'.

(4) 'Any enclosed space between... accommodation and any room in which any person... is employed to work shall be provided with effective means of ventilation'.

vide Washing Facilities Regulations 1964 Reg. 7 'Effective provision shall be made as far as reasonably practicable for ventilating rooms in which washing facilities are situated'. Note: These regs are referring to washing facilities (washbasins, etc.) as distinct from W.C.'s, and urinal accommodation.

SITING OF SANITARY ACCOMMODATION IN RELATION TO WORKROOMS —

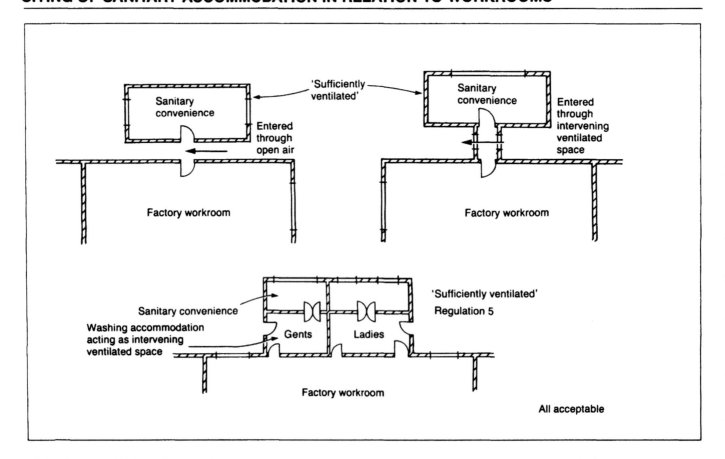

Sanitation: Ventilation of Sanitary Accom.

SCHOOLS

vide Standards for School Premises
Regulations 1972
Reg. 57(1) In every school and in all
boarding accommodation 'the washing
and sanitary accommodation shall be
soundly designed and construction, . . .
and shall . . . be reasonably
accessible. . . for whose use it was
provided'.
Note: Reg. 53(1) deals with general
ventilation requirements for 'every room'
and 53(2) specifically deals with kitchens
and 'other room(s) in which there may be
steam or fumes'. It is considered that
providing for good ventilation of sanitary
accommodation comes within the
requirement of 'sound design' (Reg.
57(1)).

BUILDING OPERATIONS AND WORKS OF ENGINEERING

vide Construction (Health and Welfare)
Regulations 1966
Reg. 14(1) 'Every sanitary convenience
shall be sufficiently ventilated, and shall
not communicate with any workroom or
messroom except through the open air or
through an intervening ventilated space'.

FACTORIES OFFICES, SHOPS AND RAILWAY PREMISES

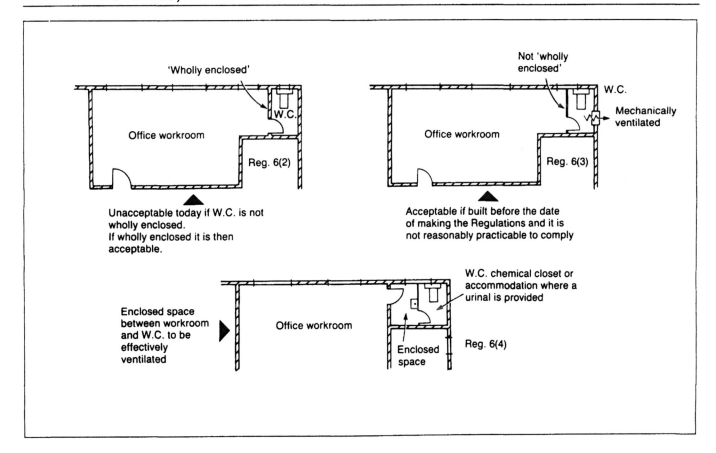

Ventilation Rates

VENTILATION RATES AND MECHANICAL VENTILATION SYSTEMS IN OUTLINE

Ventilation Requirements extracted from various authoritative references.

BS5730:1979 Code of Practice for Mechanical Ventilation and Air Conditioning in Buildings.

Information is given on the design, installation, commissioning, operation and maintenance of ventilation and air conditioning systems. The Approved Document F to the Building Regulations contains references to four clauses in Section 2:-

(i) Fresh air supply (clause 2.3.2.1) required for the dilution of odours, tobacco smoke and carbon dioxide exhaled by people. Table gives recommended quantities for various spaces.

(ii) Fire and smoke detection (clause 2.5.2.10) recommends the use of detection devices where air is recirculated. In the event of a fire, these devices can be made to shut down plant and close dampers, or discharge the smoke laden air to the outside. Suitably positioned away from the escape stairs. The detectors can be linked to alarm systems as well as dampers.

(iii) Smoke control (clause 2.5.2.11) warns of the danger of centralised ducted ventilated systems which can contribute to the spread of fire and smoke through tall buildings (stack effect). It is with this danger in mind that regulations required fans to shut down automatically when fire and smoke build up is detected. Ventilation systems should be designed with this problem in mind, with perhaps a facility to aid the removal of smoke from the building. The BS refers also to the use of ventilation systems to create positive air pressure in escape routes, such as stairways and lobbies.

(iv) Mechanical extract/natural supply (clause 3.1.1.1) described the most common form of ventilation where a fan (or more) is located on an external wall or roof, or ductwork connected to a central fan with extract points positioned in specific parts of the building complete with integral fire/smoke dampers. The exhausted air should be compensated by positioning suitable inlets into the building.

BS6465 Part 1:1984 Code of Practice for scale of provision, selection and installation of sanitary appliances. This refers to the need for adequate ventilation of all sanitary accommodation including bathrooms and kitchens. This can be achieved by windows or fanlights opening directly to the external air. It is also suggested that further ventilation can be provided by mechanical means. Where ventilation is achieved entirely by mechanical means a minimum of three air changes should be provided, preferably six or more air changes per hour should be considered.

BRE Digest 170 Ventilation of Internal Bathrooms and W.C.'s in Dwellings (Revised January 1969) suggests minimum hourly ventilation extract rates of 40 metres³/hour from a W.C. and 60 metres³/hour from W.C. in a bathroom and states, standby fan and motor are required by GLC Byelaws and Scottish Building Regulations.

BS5925:1980 Ventilation principles and designing for natural ventilation. Table 4 recommends minimum and optimum air supply rates for air conditioned spaces. For example, toilets are to be provided with 10 litres/second for each m² of floor area.

Building Regulations 1985 Approved Document F, part F1 section 2 — Mechanical Ventilation, states that where sanitary accommodation is mechanically ventilated, minimum of three air changes per hour is required. (This can be intermittent but should run for at least 15 minutes after the use of the room, or space, stops).

The Approved Document refers only to Clauses 11 to 15 of the BS (Section 3) which has a direct bearing on the subject matter applicable — i.e. dwellings, sanitary accommodation and bathrooms. The BS, however, applies to all types and size of buildings.

A.S.H.R.A.E. Guide Recommendation For groups of toilets in large air conditioned buildings — 0.01 m³/s per metre of floor area.*

I.H.V.E. Guide Natural ventilation will tend to lead to variable or reverse air flow which may contravene Regulations or Byelaws. Internal lavatories and toilets:- six to eight air changes per hour, applying to normal public toilet areas incorporating both W.C.'s and wash basins. A basis of 10 litres per second per metre² of floor area* is suggested for small or congested toilet areas comprising purely W.C.'s and urinals (*approximates to fifteen air changes per hour for single W.C. cubicle), all to be obtained by mechanical ventilation.

Manufacturers' Suggestions Xpelair — Ten to fifteen air changes per hour for lavatories Vent-Axia — Ten air changes per hour for bathrooms and toilets (minimum).

Note 1. Small rooms require a greater air change rate than larger rooms. 2. If mechanical ventilation is required, engineers generally design for six air changes per hour as the desirable minimum.

Sanitation: Ventilation of Sanitary Accom.

MECHANICAL VENTILATION SYSTEMS

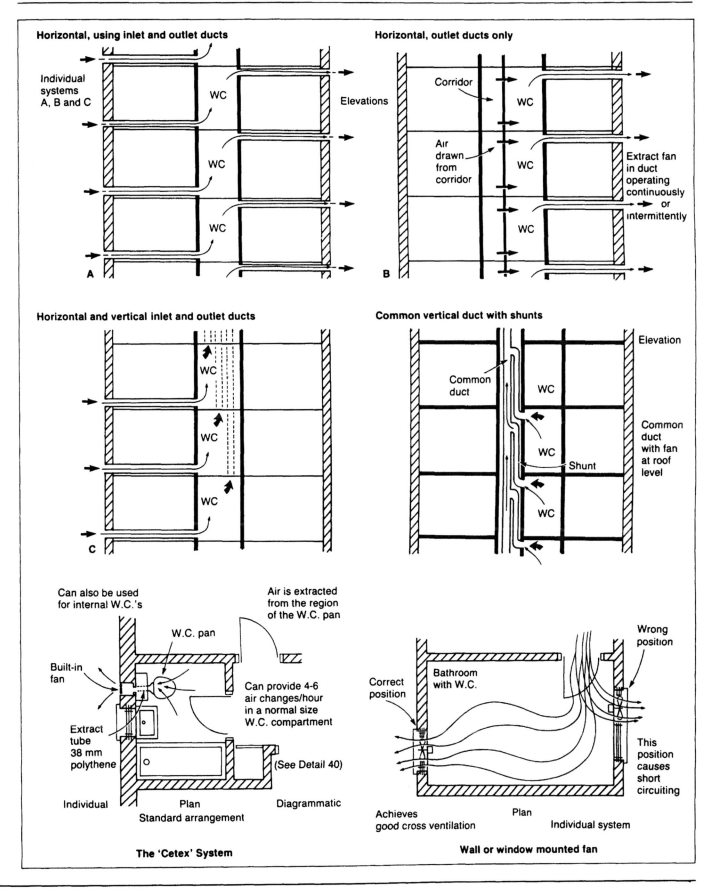

Horizontal, using inlet and outlet ducts

Individual systems A, B and C

WC

WC

WC

Elevations

A

Horizontal, outlet ducts only

Corridor

Air drawn from corridor

WC

WC

WC

Extract fan in duct operating continuously or intermittently

B

Horizontal and vertical inlet and outlet ducts

WC

WC

WC

C

Common vertical duct with shunts

Common duct

WC

WC

WC

Shunt

Elevation

Common duct with fan at roof level

Can also be used for internal W.C.'s

Air is extracted from the region of the W.C. pan

W.C. pan

Built-in fan

Extract tube 38 mm polythene

Can provide 4-6 air changes/hour in a normal size W.C. compartment

(See Detail 40)

Individual

Plan
Standard arrangement

Diagrammatic

The 'Cetex' System

Correct position

Bathroom with W.C.

Wrong position

This position causes short circuiting

Achieves good cross ventilation

Plan

Individual system

Wall or window mounted fan

FOR EXTERNAL SANITARY ACCOMMODATION

Ventaxia — window mounting 'T-series'
colour — two tone grey (3 speed) can be fitted
in single and double-glazed windows

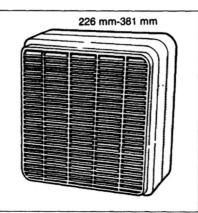

226 mm-381 mm

Size	Extract rate (medium speed)
150 dia	315 m³/hr
190 dia	395 "
230 dia	685 "
300 dia	1415 "

Roof, wall and panel models available

Standard Range
Window model — colour black or ivory can be
fitted to single glazing generally, including
fixing plates up to 10 mm thick. Single speed
intake or extract (when used with controller it
can provide variable direction of air flow and a
choice of 3 speeds).

Roof, wall and panel models available.

Sizes —	150 dia	285
	190 dia	425
	230 dia	710
	300 dia	1560

extract performance

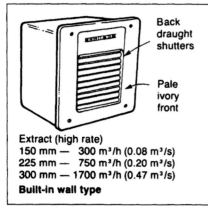

Back
draught
shutters

Pale
ivory
front

Extract (high rate)
150 mm — 300 m³/h (0.08 m³/s)
225 mm — 750 m³/h (0.20 m³/s)
300 mm — 1700 m³/h (0.47 m³/s)
Built-in wall type

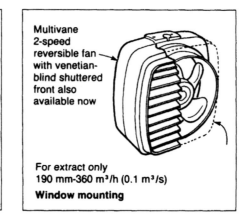

Multivane
2-speed
reversible fan
with venetian-
blind shuttered
front also
available now

For extract only
190 mm-360 m³/h (0.1 m³/s)
Window mounting

SELECTION OF SIZE

It is necessary to choose a fan or fans for
the duty to be performed. The calculation
is based on the volume of the room and
the rate of air change required.

Example
A combined bathroom and W.C.
measuring 2.14 m × 2.14 m × 2.44 m
high = 11.2 cubic metres
Number of air changes per hour = 20
Air/movement required:- 11.2 × 20 =
224 m³/h or 0.062 m³/second

Fan required e.g. Xpelair FXC6 (150 mm)
extracting 0.063 m³/sec. or Vent-Axia
(150 mm) extracting 198 m³/h or
0.06 m³/s @ low speed and 284 m³/h or
0.08 m³/s @ normal speed.

Specialist Models
— Darkroom model (4 sizes) colour —
black
— Acid/steam resistant window and roof
models
— Vehicle models (manufacturer to be
consulted)

VERTICAL FIXING

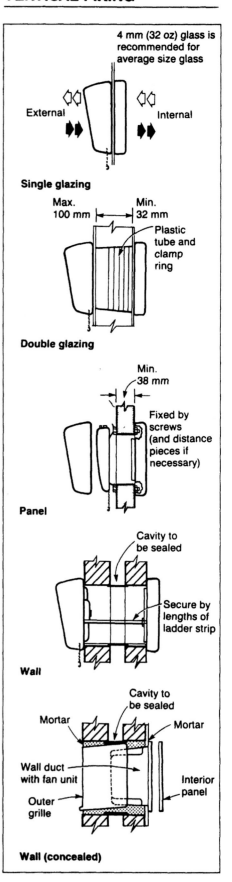

4 mm (32 oz) glass is
recommended for
average size glass

External   Internal

Single glazing

Max.   Min.
100 mm   32 mm

Plastic
tube and
clamp
ring

Double glazing

Min.
38 mm

Fixed by
screws
(and distance
pieces if
necessary)

Panel

Cavity to
be sealed

Secure by
lengths of
ladder strip

Wall

Cavity to
be sealed

Mortar   Mortar

Wall duct
with fan unit   Interior
panel

Outer
grille

Wall (concealed)

Sanitation: Ventilation of Sanitary Accom.

ROOF FIXING

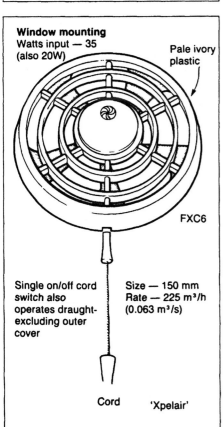

Suitably weathered

mushroom cowl

Fix in glass or panels
as for window mounted fans

Darkroom
cowl

'Xpelair'

Weathered as
necessary

'Vent-Axia'

Suitable for north lights, flat
roofs and all sloping surfaces

DETAILS OF WALL FIXING

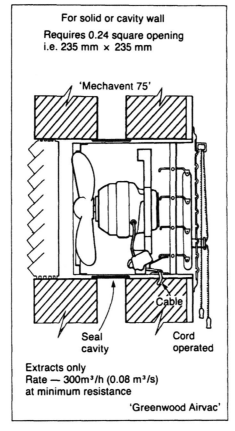

For solid or cavity wall
Requires 0.24 square opening
i.e. 235 mm × 235 mm

'Mechavent 75'

Cable

Seal
cavity

Cord
operated

Extracts only
Rate — 300m³/h (0.08 m³/s)
at minimum resistance

'Greenwood Airvac'

TYPICAL CONTROLLERS

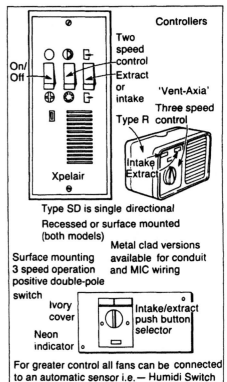

Controllers

On/
Off

Two
speed
control

Extract
or
intake

'Vent-Axia'

Three speed
control

Type R

Intake
Extract

Xpelair

Type SD is single directional

Recessed or surface mounted
(both models)

Surface mounting
3 speed operation
positive double-pole
switch

Ivory
cover

Neon
indicator

Metal clad versions
available for conduit
and MIC wiring

Intake/extract
push button
selector

For greater control all fans can be connected
to an automatic sensor i.e. — Humidi Switch
— setting range 45% to 90% R.H.
— Thermo Switch — temperature range
5°C-30°C

Window mounting
Watts input — 35
(also 20W)

Pale ivory
plastic

FXC6

Single on/off cord
switch also
operates draught-
excluding outer
cover

Size — 150 mm
Rate — 225 m³/h
(0.063 m³/s)

Cord

'Xpelair'

Wall model
For standard 225 mm solid, 280 mm cavity
or thicker walls

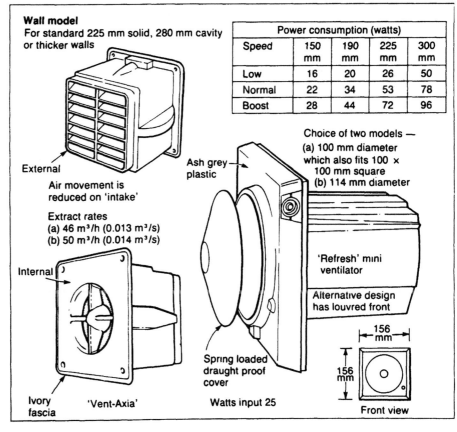

External

Air movement is
reduced on 'intake'

Extract rates
(a) 46 m³/h (0.013 m³/s)
(b) 50 m³/h (0.014 m³/s)

Internal

Ivory
fascia

'Vent-Axia'

Ash grey
plastic

Spring loaded
draught proof
cover

Watts input 25

Power consumption (watts)				
Speed	150 mm	190 mm	225 mm	300 mm
Low	16	20	26	50
Normal	22	34	53	78
Boost	28	44	72	96

Choice of two models —
(a) 100 mm diameter
which also fits 100 ×
100 mm square
(b) 114 mm diameter

'Refresh' mini
ventilator

Alternative design
has louvred front

156
mm

156
mm

Front view

'LOOVENT' TOILET VENTILATION UNIT

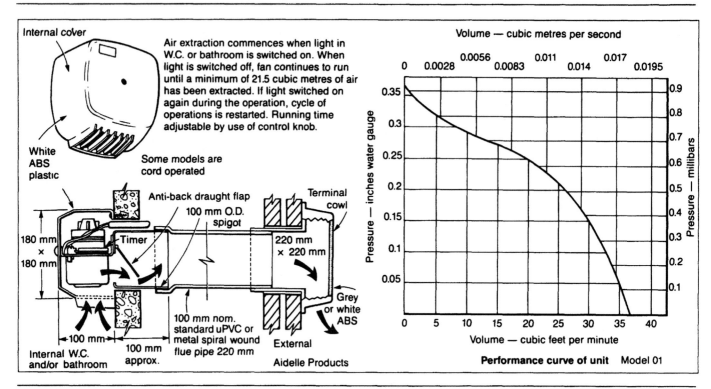

Internal cover

White ABS plastic

Air extraction commences when light in W.C. or bathroom is switched on. When light is switched off, fan continues to run until a minimum of 21.5 cubic metres of air has been extracted. If light switched on again during the operation, cycle of operations is restarted. Running time adjustable by use of control knob.

Some models are cord operated

Anti-back draught flap
100 mm O.D. spigot
Terminal cowl
220 mm × 220 mm

Timer
180 mm × 180 mm

Grey or white ABS

100 mm

Internal W.C. and/or bathroom
100 mm approx.
100 mm nom. standard uPVC or metal spiral wound flue pipe 220 mm
External

Aidelle Products

Volume — cubic metres per second

Performance curve of unit   Model 01

'MECHAVENT' TYPE PF HORIZONTAL EXTRACT SYSTEM

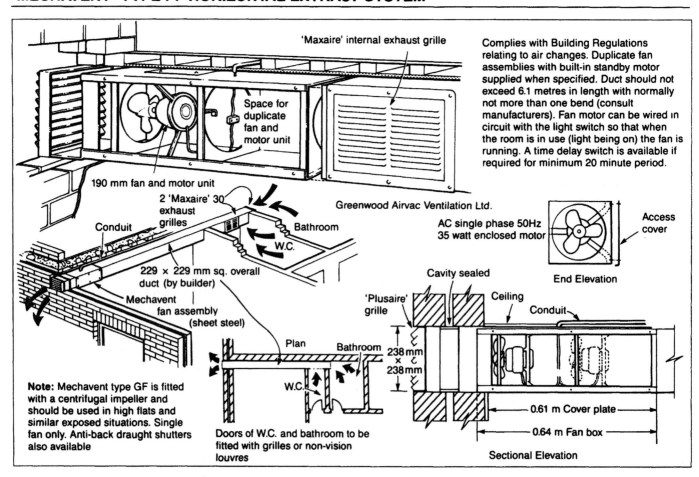

'Maxaire' internal exhaust grille

Space for duplicate fan and motor unit

190 mm fan and motor unit

2 'Maxaire' 30 exhaust grilles

Conduit

Bathroom
W.C.

229 × 229 mm sq. overall duct (by builder)

Mechavent fan assembly (sheet steel)

Plan
Bathroom
W.C.

Doors of W.C. and bathroom to be fitted with grilles or non-vision louvres

Note: Mechavent type GF is fitted with a centrifugal impeller and should be used in high flats and similar exposed situations. Single fan only. Anti-back draught shutters also available

Complies with Building Regulations relating to air changes. Duplicate fan assemblies with built-in standby motor supplied when specified. Duct should not exceed 6.1 metres in length with normally not more than one bend (consult manufacturers). Fan motor can be wired in circuit with the light switch so that when the room is in use (light being on) the fan is running. A time delay switch is available if required for minimum 20 minute period.

Greenwood Airvac Ventilation Ltd.

AC single phase 50Hz 35 watt enclosed motor

Access cover

End Elevation

Cavity sealed
'Plusaire' grille
Ceiling
Conduit

238 mm × 238 mm

0.61 m Cover plate
0.64 m Fan box

Sectional Elevation

The 'Harmony P and PD' surface mounted extract fans from Greenwood Airvac each contain a double inlet impeller designed to be self balancing for most application.

An air flow of 85 m³/W is attained against a static head of 30 N/mm² when only one intake is used the air flow will be reduced to approximately 75 per cent. The installed resistance will be dependent on the ductwork: Data for calculation of static pressure are given in the IHVE Guide for Pipes and Bends in a variety of materials.

Type PD Extract Fan
with second standby fan and motor.

— accessories available include delay timer (10-30 minutes), anti-backdraught shutter, secondary grilles, exhaust baffles and cowls etc.

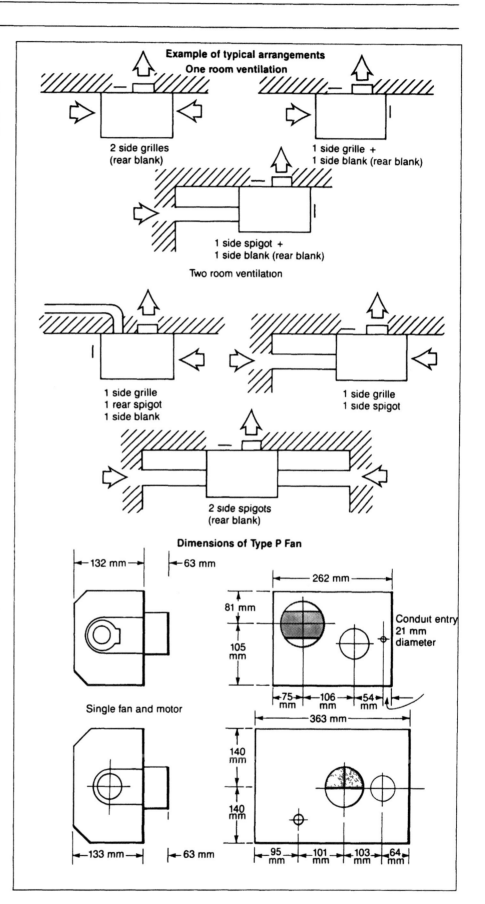

Example of typical arrangements
One room ventilation

2 side grilles
(rear blank)

1 side grille +
1 side blank (rear blank)

1 side spigot +
1 side blank (rear blank)

Two room ventilation

1 side grille
1 rear spigot
1 side blank

1 side grille
1 side spigot

2 side spigots
(rear blank)

Dimensions of Type P Fan

132 mm

63 mm

262 mm

81 mm

105 mm

Conduit entry
21 mm
diameter

75 mm — 106 mm — 54 mm

363 mm

140 mm

140 mm

Single fan and motor

133 mm

63 mm

95 mm — 101 mm — 103 mm — 64 mm

THE CETEX SYSTEM FOR INTERNAL AND EXTERNAL SANITARY ACCOMMODATION

The basic system ensures that no foul air can disperse into the room space by extracting air from the region of the W.C. pan. The basic arrangement deals with one W.C. only and can be arranged to operate during occupation of the W.C. or via a 20 minute delay switch but the latter is not absolutely necessary.

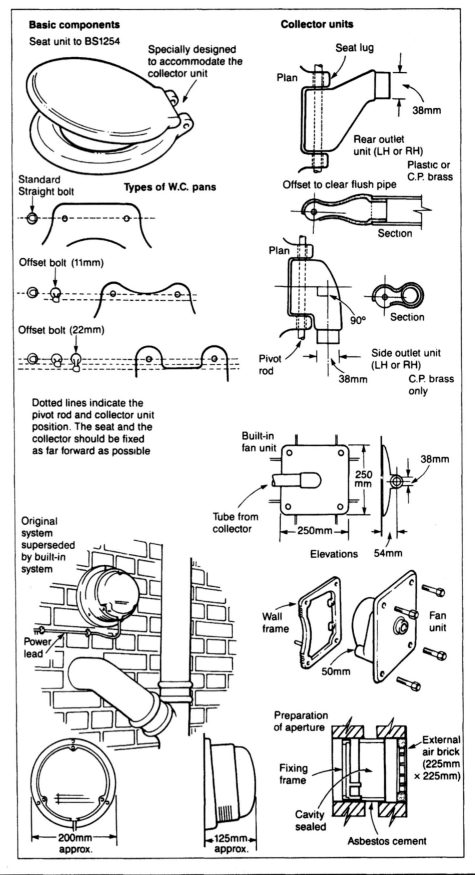

Basic components
Seat unit to BS1254

Specially designed to accommodate the collector unit

Standard Straight bolt

Types of W.C. pans

Offset bolt (11mm)

Offset bolt (22mm)

Dotted lines indicate the pivot rod and collector unit position. The seat and the collector should be fixed as far forward as possible

Original system superseded by built-in system

Power lead

200mm approx.

125mm approx.

Collector units

Seat lug

Plan

38mm

Rear outlet unit (LH or RH)

Plastic or C.P. brass

Offset to clear flush pipe

Section

Plan

90°

Section

Pivot rod

Side outlet unit (LH or RH)

38mm

C.P. brass only

Built-in fan unit

250 mm

38mm

Tube from collector

250mm

Elevations

54mm

Wall frame

Fan unit

50mm

Preparation of aperture

External air brick (225mm × 225mm)

Fixing frame

Cavity sealed

Asbestos cement

BUILT-IN VENTILATION

Normal application for single W.C.

Plan   Plan

Using 'Standard' collector

Using LH side outlet collector

Section

RH collector unit

Fan unit

Polythene tube with push-fit joints (white) Normal size 38mm

Air extraction from the region of the bowl ensures that no foul air can disperse into the room space — W.C. or bathroom

Electrical. Recommended control — silent ceiling switch located near to W.C. Earthing is essential

Flush pipe — Bends   Fan unit

Method when soil pipe is in line with W.C. pan

Using LH collector

Soil pipe

Plan

Flush pipe

Plan

Sketches are diagrammatic

A selection of arrangements

Extract to the rear

Offset to the rear

Extract tubing taken through floor

Extract tubing taken upwards

Using the built-in fan unit, the Silavent System can be installed whereever the toilet may be situated including internal locations. The fan unit should always be kept as close to the pan as possible, as the length of 38 mm tubing limits the air flow. Sufficient extra length of 50 mm tubing should be allowed to enable the fan unit to be removed for servicing. Wall frame is not required when the fan unit is fitted into partitioning, as unit can be screwed directly to the material

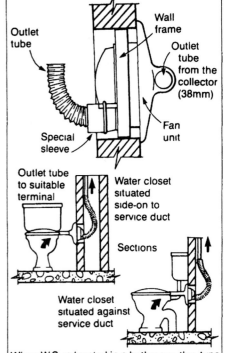

Outlet tube

Wall frame

Outlet tube from the collector (38mm)

Special sleeve

Fan unit

Outlet tube to suitable terminal

Water closet situated side-on to service duct

Sections

Water closet situated against service duct

When W.C. is located in a bathroom, this type of fan unit can be used. It has a higher air extract rate and will normally meet the Building Regulation requirement of minimum three air changes per hour

Combined bathroom and W.C. ventilation

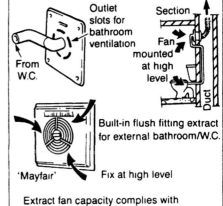

Outlet slots for bathroom ventilation

Section

Fan mounted at high level

From W.C.

Duct

Built-in flush fitting extract for external bathroom/W.C.

'Mayfair'   Fix at high level

Extract fan capacity complies with normal requirements and regulations. 50mm auxiliary inlet and separate timer available if required. Performance - 85m³/h free air. Six air changes/h for bathroom or bathroom plus adjacent W.C. External non-return flaps recommended.

Common Duct Ventilation Systems

MECHANICAL VENTILATION OF INTERNAL W.C.'s AND BATHROOMS IN HIGH RISE

Brief Notes. Common duct mechanical extract systems, operating continuously and entirely separate from any other ventilation system, should provide for a minimum extract rate of 20 metres³/hour from a bathroom with a W.C. pan (BRE Digest 170). For preference, vertical shunts as opposed to horizontal stub branches should connect the extract grilles to the common main (or subsidiary) vertical duct. The advantages claimed for the shunt system are (i) better sound attenuation between dwellings. (ii) should a fire occur, smoke is less likely to be carried from one dwelling to another. (iii) the system is often more compact.

Integration of services is of utmost importance and with careful design, ventilation ducts can be accommodated in a single service duct with drainage, rainwater, cold and hot water, fire mains, etc.

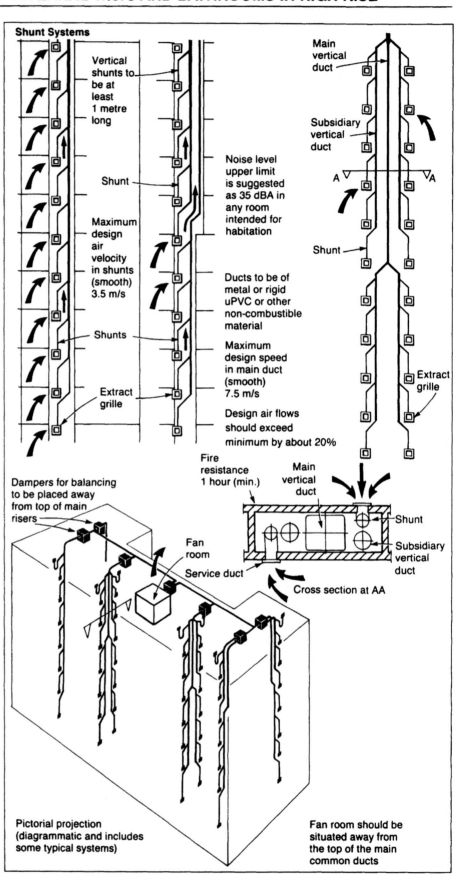

Shunt Systems

Vertical shunts to be at least 1 metre long

Shunt

Maximum design air velocity in shunts (smooth) 3.5 m/s

Shunts

Extract grille

Noise level upper limit is suggested as 35 dBA in any room intended for habitation

Ducts to be of metal or rigid uPVC or other non-combustible material

Maximum design speed in main duct (smooth) 7.5 m/s

Design air flows should exceed minimum by about 20%

Main vertical duct

Subsidiary vertical duct

Shunt

Extract grille

Main vertical duct

Fire resistance 1 hour (min.)

Shunt

Subsidiary vertical duct

Cross section at AA

Dampers for balancing to be placed away from top of main risers

Fan room

Service duct

Pictorial projection (diagrammatic and includes some typical systems)

Fan room should be situated away from the top of the main common ducts

DWELLINGS

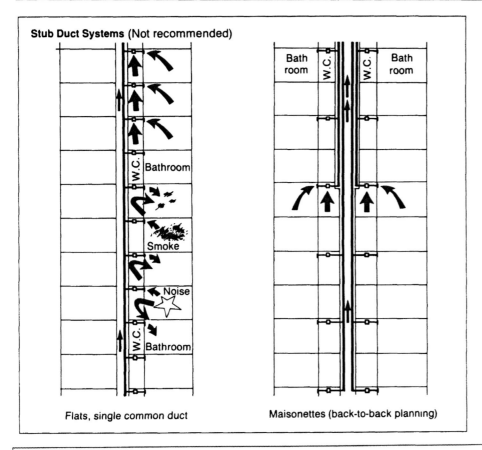

Stub Duct Systems (Not recommended)

Flats, single common duct

Maisonettes (back-to-back planning)

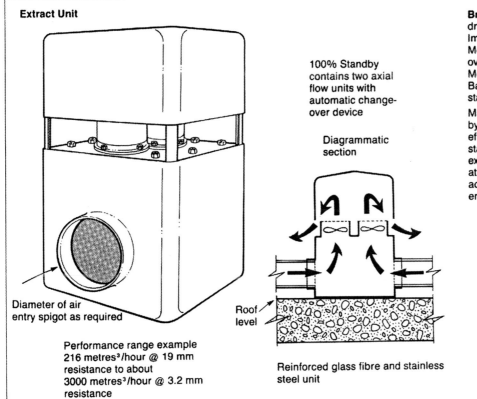

Extract Unit

Diameter of air
entry spigot as required

Performance range example
216 metres³/hour @ 19 mm
resistance to about
3000 metres³/hour @ 3.2 mm
resistance

100% Standby
contains two axial
flow units with
automatic change-
over device

Diagrammatic
section

Roof
level

Reinforced glass fibre and stainless
steel unit

Brief Specification. Fans — two direct
drive, aerofoil section axial flow type.
Impellers — moulded phenolic resin.
Motors — Totally enclosed, non-
overloading at any duty.
Mounting shutter — anti-vibration pads.
Backdraught shutter — automatic in
stainless steel with nylon bearings.

Manufactured from materials unaffected
by corrosion conditions. Has high
efficiency output with maximum possible
stability of air volume regardless of all
exterior wind conditions. Suction
attenuation chamber and outlet cowl is
acoustically treated to ensure low sound
emission.

Systems in Outline

GENERAL PRINCIPLES

Soil and waste pipe systems should comprise the minimum of pipework necessary to carry away the foul and dirty water from the building quickly, reliably, quietly and with freedom from nuisance or risk of injury to health.

The system should satisfy the following requirements:-

(1) Efficient and speedy removal of excremental matter and urine plus certain other liquids and solids, without leakage.

(2) the prevention of ingress of foul 'air' to the building whilst providing for their escape from the system in a 'safe' position.

(3) The adequate and easy access to the interior of the pipe for the clearance of obstructions.

(4) Protection against extremes of temperature (freezing conditions and close proximity to hot pipes) by careful siting.

(5) The prevention of external and internal corrosion and erosion by correct choice of materials and protection.

(6) Correct design to limit siphonage (if any) to an acceptable standard; and to avoid deposition; and damage and proneness to obstruction.

(7) In areas where a combined system of drainage is permitted, it may be advantageous to connect roof rainwater outlets directly to discharge pipes. However, BS5572:1978 points out the risk of flooding, should a blockage occur in the discharge stack or underground drain serving multi-storey buildings (i.e. 30 storeys). It is also stated that even with small continuous flows of rainwater, excessive pressure fluctuations can be caused.

(8) Economy and good design are essentials. Both are aided by compact grouping of sanitary appliances in both horizontal and vertical directions.

BUILDING REGULATIONS 1985

Building Regulations 1985, Approved Document H1 "Sanitary Pipework and Drainage" refers to the following system of pipework:-

(i) single stack system
(ii) modified single stack system
(iii) ventilated system

BS5572:1978 gives the following descriptions of these systems:

(i) Single stack system — is used in situations where close grouping of appliances make it practicable to provide branch discharge pipes without the need for branch ventilating pipes, but only where the discharge stack is large enough to limit pressure fluctuations without the need for a ventilating stack.

(ii) Modified single stack system — providing ventilating pipework extended to the atmosphere or connected to a ventilated stack, can be used where the disposition of appliances on a branch discharge pipe could cause loss of their trap seals. The ventilating stack need not be connected directly to the discharge stack and can be smaller in diameter than that required for a ventilated stack system.

(iii) Ventilated system — used where there are a large number of sanitary appliances in ranges, or where they have to be widely dispersed and it is impracticable to provide discharge stack(s) in close proximity to the appliances. Trap seals are safeguarded by extending the discharge and ventilating stacks to atmosphere and by providing individual branch ventilating pipes.

(iv) Ventilated stack system (not described in Building Regulations) — this system is used in situations similar to those described in (i) except trap seals are safeguarded by extending the stack(s) to the atmosphere and by cross-connecting the ventilating stack to the discharge stack.

Sanitation: Sanitary Pipework Above Ground

Single stack system

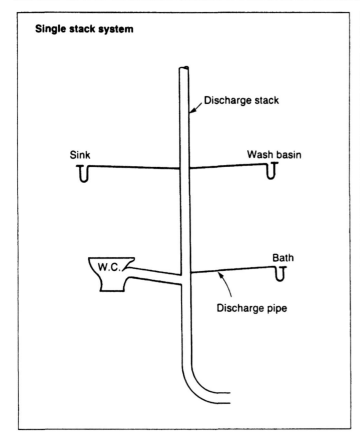

Discharge stack

Sink

Wash basin

W.C.

Bath

Discharge pipe

Modified single stack system

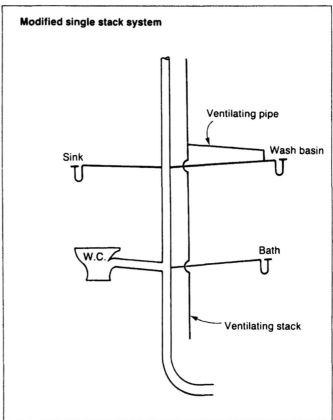

Ventilating pipe

Sink

Wash basin

W.C.

Bath

Ventilating stack

Ventilated system (single appliances)

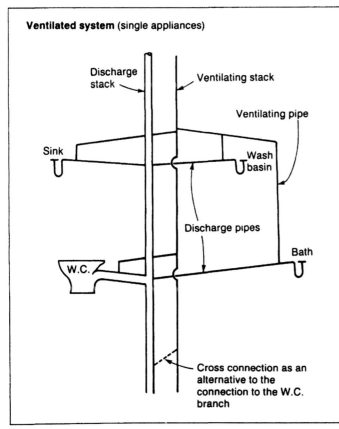

Discharge stack

Ventilating stack

Ventilating pipe

Sink

Wash basin

Discharge pipes

W.C.

Bath

Cross connection as an alternative to the connection to the W.C. branch

Ventilated stack system

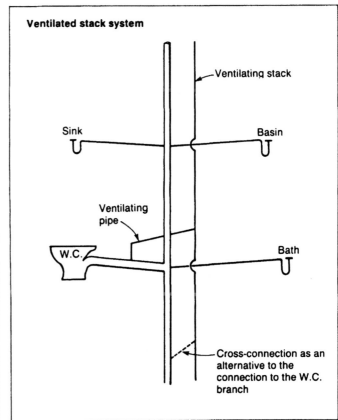

Ventilating stack

Sink

Basin

Ventilating pipe

W.C.

Bath

Cross-connection as an alternative to the connection to the W.C. branch

BS5572:1968 SANITARY PIPEWORK ABOVE GROUND

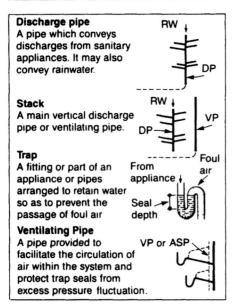

Discharge pipe
A pipe which conveys discharges from sanitary appliances. It may also convey rainwater.

Stack
A main vertical discharge pipe or ventilating pipe.

Trap
A fitting or part of an appliance or pipes arranged to retain water so as to prevent the passage of foul air

Ventilating Pipe
A pipe provided to facilitate the circulation of air within the system and protect trap seals from excess pressure fluctuation.

WATER SEALS IN TRAPS

Approved Document H1 of the Building Regulations 1985 stipulates the following:

(a) All points of discharge into the system should be fitted with a water seal (trap) to prevent foul air from entering the building under working conditions.

(b) Minimum trap sizes and seal depths are given for sanitary appliances which are most used (see table 1).

Table 1

Appliance	Diameter of trap (mm)	Depth of seal (mm)
Wash basin Bidet	32	75
Sink Bath Shower Food waste disposal Urinal Bowl	40	75
WC pan	minimum 75	50

(c) Ventilation — to prevent the water seal from being broken by the pressures which can develop in the system, the branch discharge pipes should be designed as described in paragraphs 1.5 and 1.21 of the Approved Document.

(d) Access for clearing blockages — if a trap forms part of an appliance, the appliance should be removable. All other traps should be fitted directly after the appliance, and should be removable or be fitted with a cleaning eye.

BRANCH DISCHARGE VENTILATING PIPES AND STACKS

General requirements for plumbing systems as stated in the Approved Document H1 to the Building Regs 1985.

Section 1.12 — sizes of branch pipes. Pipes serving a single appliance should have at least the same diameter as the appliance trap (see Table 1), if a pipe serves more than one appliance and is unventilated, the diameter should be at least the size shown in Table 2.

Table 2

Appliance	Max. No. to be connected	OR Max. length of branch	Min. size of pipes	Gradient limits per full m Min Max
WC	8	15	100	9 to 90
Urinals bowls: stalls:	5 6	* *	50 65	18 to 90 18 to 90
Wash basins	4	4 (no bends)	50	18 to 45

* No limitation as regards venting, but should be as short as possible.

Section 1.24 — minimum sizes of discharge stacks. Stacks should have at least the diameter shown in Table 3 and should reduce in the direction of the flow. Stacks serving urinals should not be less than 50 mm, and stacks serving closets not less than 75 mm.

Table 3 — Maximum capacities for discharge stacks

Stack size (mm)	Max. capacity (litres/sec)
50	1.2*
65	2.1*
75	3.4+
90	5.3
100	7.2

+ Not more than one siphonic wc with 75 mm outlet   * No wc's

Section 1.13 — bends. Bends in branch pipes should be avoided if possible, where they cannot they should have as large a radius as possible. Pipes of 65 mm or less should have a centre line radius of at least 75 mm. Junctions on branch pipes should be made with a sweep of 25 mm radius or at 45 deg. (BS5572 Amendment 4202 dated June 1983 gives a minimum radius from branch connections of 50 mm).

Connection of branch pipes of 75 mm diameter or more to the stack should be made with a sweep of 50 mm radius or at 45 deg.

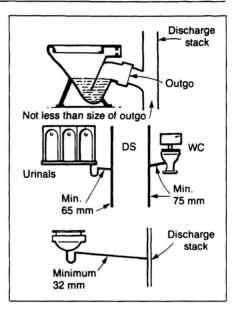

Section 1.31 — pipe supports. Pipes should be firmly supported without restricting thermal movement.

Section 1.32 — watertightness. The

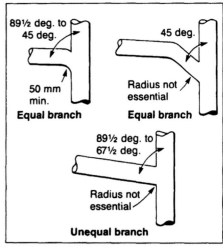

installation should be capable of withstanding an air or smoke test of positive pressure of at least 38 mm water gauge for at least 3 minutes. During this time, every trap should maintain a water seal of at least 25 mm.

Sections 1.21 and 1.30 — access for clearing blockages to lengths of discharge pipes and ventilating pipes. Rodding points are to be provided to lengths which cannot be reached by removing traps. The top of a ventilating pipe may provide access.

Minimum size of wash basin waste pipe — ID (Internal diameter) of waste pipe shall be not less than 32 mm if it serves a (wash) basin.

Materials — Any stack discharge pipe or ventilating pipe shall be composed of suitable materials of adequate strength and durability.

Materials for pipes, joints and fittings — The following materials can be used for sanitary pipework. Different metals should be separated by non-metallic material to prevent electrolytic corrosion.

Pipes — Cast iron to BS416. Copper to BS864 and BS2871. Galvanised steel to BS3868. UPVC to BS4514. Polypropylene to BS5254. **Traps** — Copper to BS1184. Plastics to BS3943

Further requirements for discharge stacks — Approved Document H1 of the Building Regulations 1985 gives the following information.

Section 1.27. Ventilating pipes open to the outside air should finish at least 900 mm above any opening into the building within 3 m and should be finished with a cage or other cover which does not restrict the flow of air.

Section 1.25. Within certain limitations, an unvented stack (Stub Stack) can be used, provided the stack connects into a ventilated discharge stack or a drain, and no branch into the stack is more than 2 m above the invert of the connection or drain and no branch serving a closet is more than 1.5 m from the crown of the closet trap to the invert of the connection or drain.

Section 1.26. Where a stub stack is used, there may still be parts of the system which should be ventilated.

Section 1.29. Discharge stacks may terminate inside a building when fitted with air admittance valves. Where these are used, they should not adversely affect the ventilation necessary to the below ground system which is normally

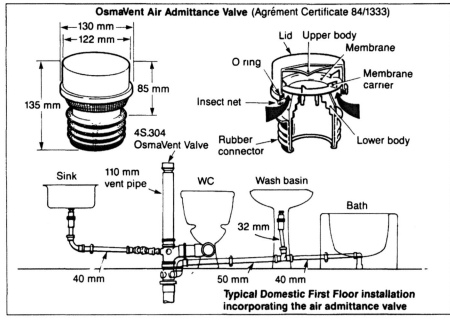

OsmaVent Air Admittance Valve (Agrément Certificate 84/1333)

130 mm
122 mm
85 mm
135 mm
4S.304 OsmaVent Valve

Lid Upper body
Membrane
O ring
Membrane carrier
Insect net
Rubber connector
Lower body

Sink
110 mm vent pipe
WC
Wash basin
Bath
32 mm
40 mm
50 mm
40 mm

Typical Domestic First Floor installation incorporating the air admittance valve

provided by the open stacks of the sanitary pipework. Only an air admittance valve which carries a British Board of Agrément Certificate should be used and the conditions of use should meet with the terms of the certificate.

This type of unit avoids the need for a soil stack to penetrate the roof construction. It is suitable for use on buildings up to five storeys, with two groups of appliances per floor (in domestic buildings a group of appliances is one wc, one wash basin, one sink and one bath and/or shower).

No provision is made in Approved Document H1 for the discharge of rainwater into the sanitary pipework system. Reference should be made to Approved Document H3.

Bends — If bends are necessary, they must not form an acute angle but have the largest practicable radius. There shall be no change in cross-section of pipe throughout the bend.

Supports — Shall be adequately supported without restraining thermal movement, with any fitting giving support being securely attached to the building.

Testing — (watertightness) Shall be capable of withstanding smoke or air test for a minimum period of 3 minutes at a pressure equivalent to a head of not less than 38 mm of water.

Internal access — Shall have such means of access as are necessary to permit internal cleansing.

Public Health Act 1936 and 1961 — Section 40 — Provision as to soil pipes

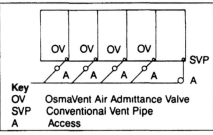

Key
OV — OsmaVent Air Admittance Valve
SVP — Conventional Vent Pipe
A — Access

and ventilating pipes.

(1) No pipe for conveying rainwater from a roof shall be used for the purpose of conveying the soil or drainage from any sanitary appliance. (Note: This does not prevent the opposite being done, viz: soil pipe being used to convey roofwater).

(2) The soil pipe from every WC shall be properly ventilated.

(3) No pipe for conveying surface water from any premises (e.g. an ordinary rainwater pipe) shall be permitted to act as a vent to any drain etc. conveying foul water. (Note: this does not prevent a proper drain ventilating pipe conveying roof water).

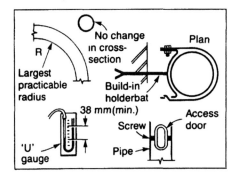

No change in cross-section
Plan
R
Largest practicable radius
Build-in holderbat
38 mm(min.)
'U' gauge
Screw
Pipe
Access door

Stub stack installation
Diagram 5 of Approved Document H1

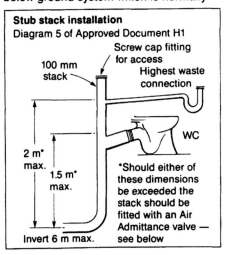

100 mm stack
Screw cap fitting for access
Highest waste connection
WC
2 m* max.
1.5 m* max.
Invert 6 m max.

*Should either of these dimensions be exceeded the stack should be fitted with an Air Admittance valve — see below

References
VP — Vent pipe
WP — Waste pipe
SP — Soil pipe
DP — Discharge pipe
ASP — Anti-siphon pipe
RW — Rainwater
WC — Water closet
WB — Lavatory (wash) basin
Note
Wash basin is preferred term
Lavatory basin non preferred (BS4118)

41

Statutory Requirements (2)

SOIL APPLIANCES

Note: Waste disposal units do not have integral trap and therefore should have a tubular trap fitted to the outgo.

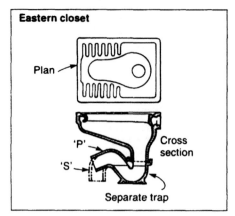
Eastern closet — Plan, Cross section, 'P', 'S', Separate trap

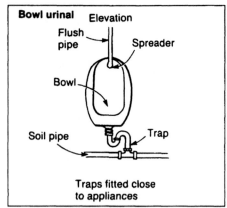
Bowl urinal — Elevation, Flush pipe, Spreader, Bowl, Soil pipe, Trap. Traps fitted close to appliances

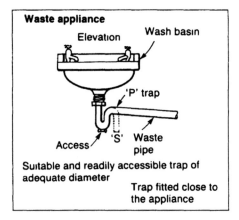
Waste appliance — Elevation, Wash basin, 'P' trap, 'S', Access, Waste pipe. Suitable and readily accessible trap of adequate diameter. Trap fitted close to the appliance

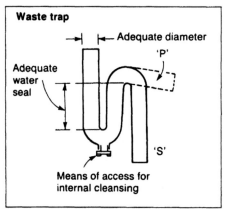
Waste trap — Adequate diameter, 'P', Adequate water seal, 'S', Means of access for internal cleansing

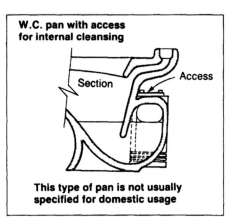
W.C. pan with access for internal cleansing — Section, Access. This type of pan is not usually specified for domestic usage

SANITARY APPLIANCES

with integral trap

Note: Any inlet to a drain other than a SP, WP or VP to be effectively trapped.

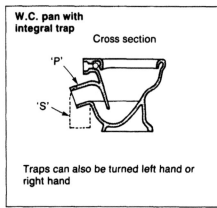
W.C. pan with integral trap — Cross section, 'P', 'S'. Traps can also be turned left hand or right hand

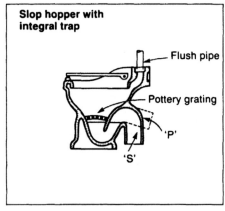
Slop hopper with integral trap — Flush pipe, Pottery grating, 'P', 'S'

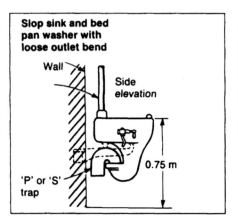
Slop sink and bed pan washer with loose outlet bend — Wall, Side elevation, 0.75 m, 'P' or 'S' trap

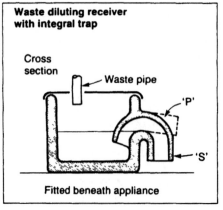
Waste diluting receiver with integral trap — Cross section, Waste pipe, 'P', 'S'. Fitted beneath appliance

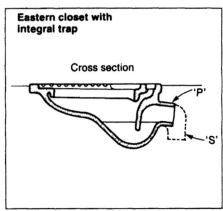
Eastern closet with integral trap — Cross section, 'P', 'S'

Sanitation: Sanitary Pipework Above Ground

Lavatory basins

Lavatory basins fixed in a range and discharging to a common waste pipe
Range of lavatory basins

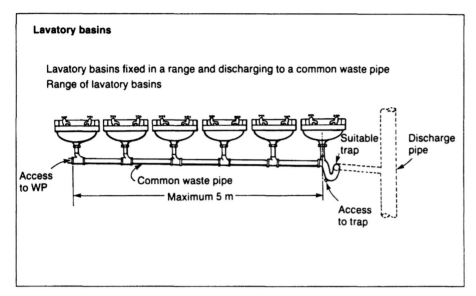

Discharge stacks

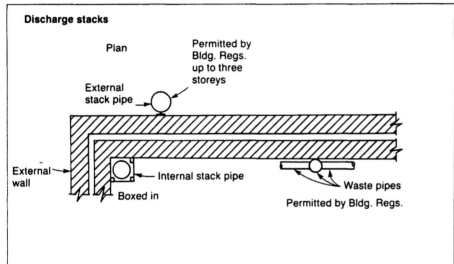

Section 1.23 of Approved Document H1 of the Building Regulations 1985 — Discharge stacks — stacks should have no offset in any part carrying foul water (wet part below the highest branch) and should be run inside the building if it has more than three storeys.

Back inlet trapped gully (BS65)

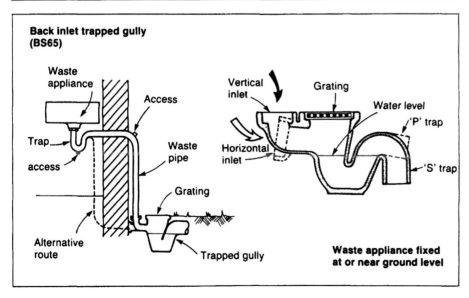

Section 1.11 of the Approved Document states: a branch pipe should only discharge to a gully between the grating and the top level of the waterseal.

DEFINITION

A fitting or part of an appliance or pipe arranged to retain water acting as a seal to prevent the passage of foul air. The water in a trap acting as a barrier to the passage of air through the trap is known as the water seal and is the depth of water which would have to be removed from a fully charged trap before air could pass through the trap.

General notes from BS5572:1978 Code of Practice Sanitary Pipework 'The entry of foul air from the drainage system into the building is prevented by the installation of suitable traps which should be self-cleansing. A trap which is not an integral part of an appliance should be attached to and immediately beneath its outlet and the bore of the trap should be smooth and uniform throughout. All traps should be accessible and be provided with adequate means of cleaning. There is advantage in providing traps which are capable of being readily removed or dismantled.'

Depths of water seals as recommended by BS5572
Water closets — minimum 50 mm
Other appliances — minimum 75 mm for traps up to and including 50 mm diameter.
Minimum 50 mm for traps over 50 mm diameter.

Some trap types
'P' — vertical inlet, and outlet inclined slightly below horizontal.
'Q' = vertical inlet, and outlet inclined at angle of about 45 deg.
'S' = vertical inlet, and outlet parallel but not in line with the inlet.
Deep seal = 75 mm (3″) or more in depth.

Minimum internal trap diameter	
Domestic appliances	Minimum internal diameter mm
Wash basin	32
Bidet	32
Sink	40
Bath	40
Shower bath tray	40
Wash tub	50
Kitchen waste disposal unit (Note: essential to fit tubular trap)	40
Non-domestic appliances	
Drinking fountain	32
Bar well	32
Hotel or restaurant sink	40
Urinal bowl	40
Urinal stall (1-6)*	50
Waste food disposal unit (Note: essential to fit tubular trap) Industrial type	50
Sanitary towel macerator	50

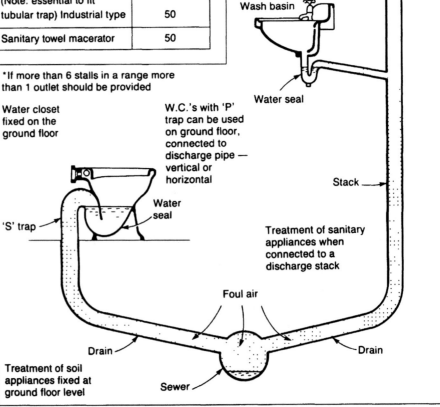

*If more than 6 stalls in a range more than 1 outlet should be provided

Water closet fixed on the ground floor

W.C.'s with 'P' trap can be used on ground floor, connected to discharge pipe — vertical or horizontal

Treatment of sanitary appliances when connected to a discharge stack

Treatment of soil appliances fixed at ground floor level

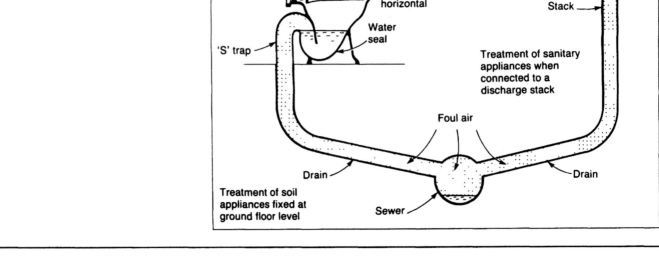

Sanitation: Sanitary Pipework Above Ground

TRAPS

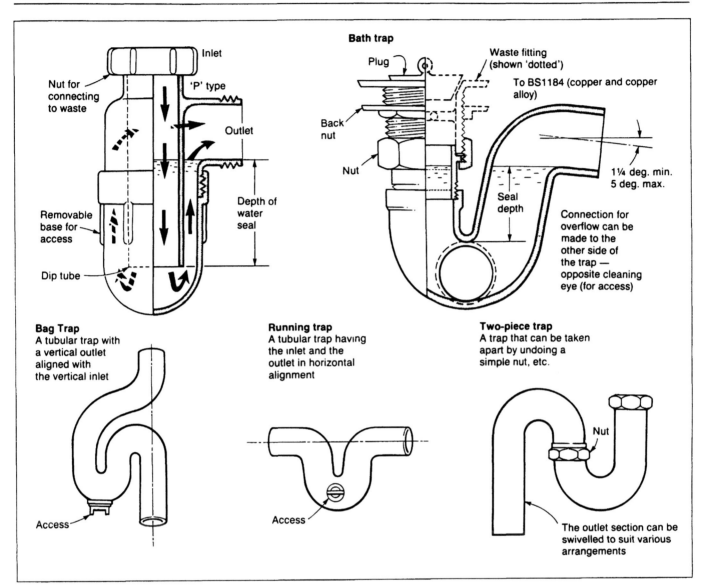

Bath trap

Bag Trap
A tubular trap with a vertical outlet aligned with the vertical inlet

Running trap
A tubular trap having the inlet and the outlet in horizontal alignment

Two-piece trap
A trap that can be taken apart by undoing a simple nut, etc.

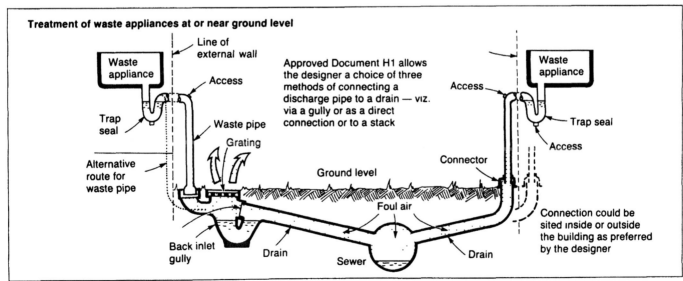

Treatment of waste appliances at or near ground level

Approved Document H1 allows the designer a choice of three methods of connecting a discharge pipe to a drain — viz. via a gully or as a direct connection or to a stack

Traps (2)

RESEALING TRAPS

Definition. A trap designed to retain an effective water seal after relieving excessive pressure fluctuations either at the inlet or the outlet of the trap. (BS4118)

Illustrating the effect of suction (negative pressure) at outlet

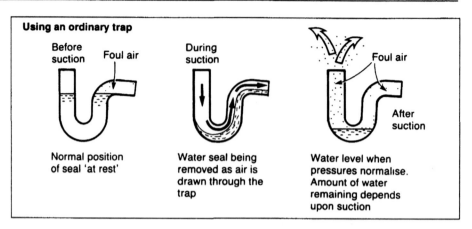

Using an ordinary trap

Before suction — Foul air

During suction

Foul air — After suction

Normal position of seal 'at rest'

Water seal being removed as air is drawn through the trap

Water level when pressures normalise. Amount of water remaining depends upon suction

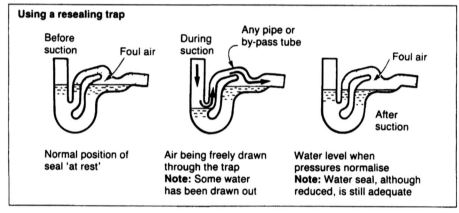

Using a resealing trap

Before suction — Foul air

During suction — Any pipe or by-pass tube

Foul air — After suction

Normal position of seal 'at rest'

Air being freely drawn through the trap
Note: Some water has been drawn out

Water level when pressures normalise
Note: Water seal, although reduced, is still adequate

SOME BASIC FORMS OF RESEALING TRAPS

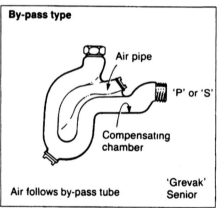

By-pass type

Air pipe

'P' or 'S'

Compensating chamber

Air follows by-pass tube

'Grevak' Senior

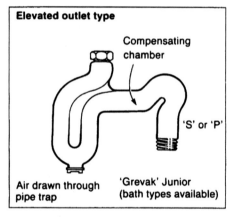

Elevated outlet type

Compensating chamber

'S' or 'P'

Air drawn through pipe trap

'Grevak' Junior (bath types available)

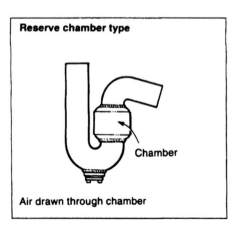

Reserve chamber type

Chamber

Air drawn through chamber

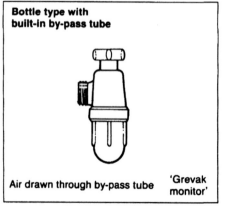

Bottle type with built-in by-pass tube

Air drawn through by-pass tube

'Grevak monitor'

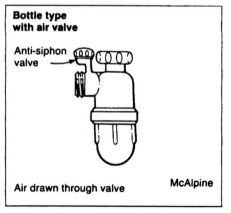

Bottle type with air valve

Anti-siphon valve

Air drawn through valve

McAlpine

Sanitation: Sanitary Pipework Above Ground

THE 'GREVAK MONITOR'

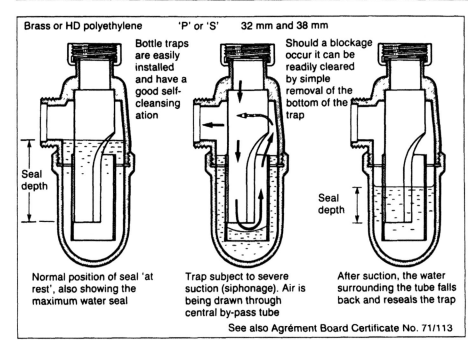

Brass or HD polyethylene   'P' or 'S'   32 mm and 38 mm

Bottle traps are easily installed and have a good self-cleansing ation

Should a blockage occur it can be readily cleared by simple removal of the bottom of the trap

Seal depth

Seal depth

Normal position of seal 'at rest', also showing the maximum water seal

Trap subject to severe suction (siphonage). Air is being drawn through central by-pass tube

After suction, the water surrounding the tube falls back and reseals the trap

See also Agrément Board Certificate No. 71/113

Graph showing performance of 32 mm 'monitor' trap compared with a standard 75 mm deep seal trap

Extracted from literature of Greenwood & Hughes

Graph shows depth of seal remaining after six test applications of suction induced by the stated flow held for 15 seconds. (Note: Seal not replenished between tests)

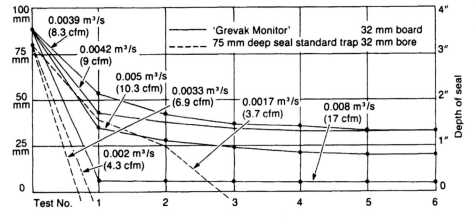

—— 'Grevak Monitor'   32 mm board
– – – – 75 mm deep seal standard trap 32 mm bore

0.0039 m³/s (8.3 cfm)
0.0042 m³/s (9 cfm)
0.005 m³/s (10.3 cfm)
0.0033 m³/s (6.9 cfm)
0.0017 m³/s (3.7 cfm)
0.008 m³/s (17 cfm)
0.002 m³/s (4.3 cfm)

Test No.   1   2   3   4   5   6

Depth of seal

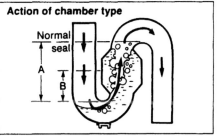

Valve detail of the McAlpine 'Silent' trap

Plastic cap
Steel valve lifts off steel seating under suction
Path or air
Trap outlet pipe wall
Suction

Action of chamber type

Normal seal
A
B

Operation. Should there be a tendency for siphonage (suction), the air inside the trap is drawn out and the valve simultaneously rises, thus neutralising the pressure. As soon as the action is finished, the valve closes air and watertight (see also BRE Report 1911).

Action of Chamber Type. Under suction, air is drawn through the trap when inlet level falls. 'B' shows seal depth after suction has taken place. This is sufficient to form an effective seal.

Extracts from Test Reports
'The performance graph shows that the 'Grevak Monitor' will even maintain its seal against suction induced by a flow rate as high as 8.1 l/s in the plumbing system' (from manufacturers' literature, based on independent test results).
'In the siphonage tests, the traps were reasonably quiet in operation and an adequate seal was maintained under conditions more severe than are likely to occur in a normal plumbing system. A seal greater than 37 mm was maintained when using a 32 mm waste pipe up to 6 m long and at slopes up to 5 deg.' (BRE Report No. 2223).
'Hyraulic tests indicate they retain an effective water seal in conditions of self-siphonage and induced siphonage in excess of those associated with normal good practice.' (Agrément Certificate No. 75/279) 1.4.1978

Other References. BRE Digest 238:1981 makes the following observation on the use of re-sealing traps: "The use of re-sealing traps with wash basins to avoid excess water seal loss from self or induced siphonage in the unventilated discharge branch pipes. These traps have been proved in practice, but they may be noisy at the end of the discharge and less efficient in re-sealing if they fill with deposits. The noise should not be a problem if their use is restricted to a single wash basin or a range of basins in a single toilet room. They should not be used to relieve the effects of excessive air pressure fluctuations in the discharge stack, through undersizing for example, nor should they be used on appliances likely to produce heavy deposition, such as sinks."

Overflows and Wastes

OVERFLOWS

Definitions. Waste appliance —
'includes a slipper bath, lavatory basin, bidet, domestic sink, cleaner's bucket sink, drinking fountain, shower tray, wash fountain, washing trough and wash tub.'
Overflow pipe (BS.4118) — 'a pipe connected to a . . . sanitary appliance . . . to discharge excess water.'
Overflow (BS4118) — 'That part of a . . . sanitary appliance . . . through which overflow is intended to take place'.

Notes from British Standards

BS5997   'Overflows should be adequate to deal with the maximum inflow. Wastes and overflows should be screened against the entry of anything likely to cause obstruction.'

CP99   'An overflow pipe from a bath should be connected to the waste trap below its waterline, but otherwise there should be no dips in overflow pipes which might trap water. A flap* on the outlet end of an overflow pipe is deprecated because it may become fixed by freezing, but the turning down of the outlet end . . . is advantageous.'
*BS5572 Overflow pipes not included.

EXAMPLES OF OVERFLOWS

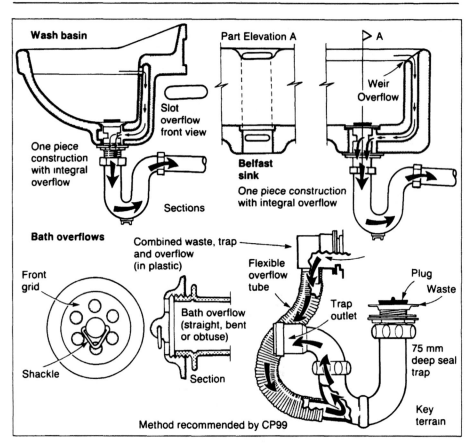

Wash basin — One piece construction with integral overflow — Slot overflow front view — Sections

Part Elevation A — A — Weir Overflow

Belfast sink — One piece construction with integral overflow

Bath overflows — Front grid — Shackle — Bath overflow (straight, bent or obtuse) — Section

Combined waste, trap and overflow (in plastic) — Flexible overflow tube — Trap outlet — Plug — Waste — 75 mm deep seal trap — Key terrain

Method recommended by CP99

BS6465: Part 1 Sanitary Appliances
Baths '. . . shold be fitted with an overflow pipe, the ID of which should be not less than 25 mm (1").'
Footbaths '. . . should be provided with an overflow.'
Washbasins '. . . should be made in one piece including a combined overflow' . . . designed to facilitate cleaning. The overflow should be of the open weir type with removable grating, or of the slot type.'
Shower trays 'Where shower trays are used as footbaths and are provided with an outlet plug, there should be an overflow.'
Domestic sink 'The overflow is usually combined with the waste and should be of the open weir type.'
Veg. sinks '. . . standing waste outlet.'
Wash tubs '. . . open weir overflow.'

British Standards (examples)
BS1188 Wash basins (Clay) 'An overflow shall be provided having a horizontal dimension not longer than 63.5 mm and an area of not less than 8 cm². Shall be of one piece construction.'

BS1189 Baths (CI) 'Overflow hole to be provided, unless otherwise ordered when a 6.4 mm hole for a chain stay shall be provided .'

BS1244 Metal sinks 'Unless otherwise ordered, sinks shall have an overflow not less than 6.45 cm² in area . . . and capable of being connected to the waste outlet.'

BS1206 Clay sinks 'Shall be of one-piece construction, including a combined overflow. Plain sinks shall have overflows of the weir type. Shelf type to have slot type.'

Definition from BS4118 'A fitting usually of brass, the function of which is to couple together in a watertight manner a waste appliance or a urinal, and the waste or soil pipe which conducts from the appliance.'

Sanitation: Sanitary Pipework Above Ground

WASTES

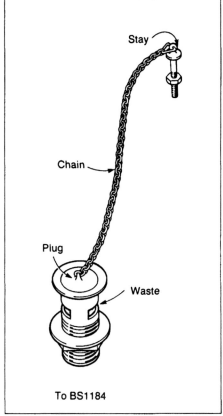

Stay

Chain

Plug

Waste

To BS1184

'Pop-up' waste

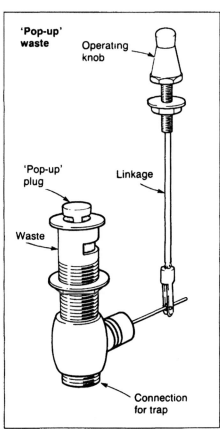

Operating knob

'Pop-up' plug

Linkage

Waste

Connection for trap

'Lift up' waste

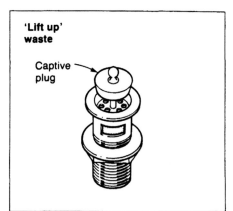

Captive plug

Standard Waste for sanitary appliances

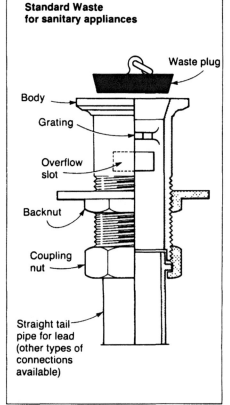

Waste plug

Body

Grating

Overflow slot

Backnut

Coupling nut

Straight tail pipe for lead (other types of connections available)

Domed outlet for channel grating

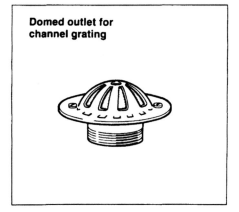

Skeleton sink waste
Used normally with lead traps and fireclay sinks

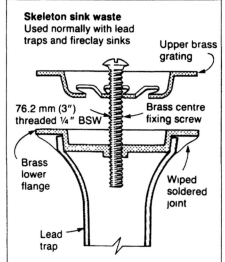

Upper brass grating

76.2 mm (3") threaded ¼" BSW

Brass centre fixing screw

Brass lower flange

Wiped soldered joint

Lead trap

Standing waste overflow

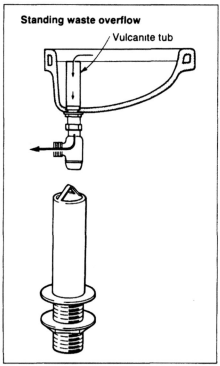

Vulcanite tub

Types of Overflows (BS4118)
Slot overflow — '. . . inlet in the form of a slot or series of slots'
Standing overflow — '. . . vertical tube standing in the waste outlet acting as a plug and overflow'
Weir overflow — '. . . inlet in the form of a weir, arranged so that the water way leading from it is accessible for cleaning through most of its length'

BUILDING REGULATIONS

Water Seals in Traps. Part H of Schedule 1 to the Building Regulations 1985 states that any system which carries foul water from appliances within the building to a foul water outfall shall be adequate. "Acceptable Level of Performance" is referred to in the Approved Document H1... systems should prevent foul air from the drainage system from entering the building under working conditions.

SEALING

Importance of Sealing. For any drainage system (above or below ground) to operate satisfactorily, the traps to the various appliances must remain sealed in all conditions of use, otherwise there is a risk of odours (e.g. drain air) contaminating rooms in different parts of the Building. A seal will be broken if the pressure changes in the branch pipe are of sufficient size and duration to overcome the head of water in the trap itself. One way of limiting the variations in air pressure during discharge is to provide an extensive system of vent piping with the object of equalizing pressures throughout the system. Research has shown that this may also be achieved by appropriate design of the drainage pipework itself, thus avoiding the need for expensive venting and the single stack drainage recommended by the BRE, is based on this approach. Another way of ensuring sufficient seal retention in traps to certain waste appliances is to use special resealing traps, but opinions on their performance differ widely.

Causes of Unsealing. Research has shown that the main causes of loss of seal are self-siphonage due to flow in the branch pipe connecting an appliance to the stack, and induced siphonage and back pressure due to flow of water down the stack. Other causes of loss of water from seals can be attributed to evaporation and wind effect across the top of tall stacks. Simple leakage from the trap is an obvious cause easily remedied and capillary attraction must be a rare cause.

Ventilating pipe (ASP) connected to (or close to) the outlet side of a trap seal is one way of safeguarding against unsealing.

SELF-SIPHONAGE

Due to their shape, wash basins are prone to self-siphonage (bidets similarly) but not baths, sinks, W.C.'s etc.

Self-siphonage is defined as the extraction of water from a trap by siphonage set up by the momentum of the discharge from the appliance to which the trap is attached (BS4118).

Even though flat-bottomed appliances may be unsealed by the main discharge, resealing is accomplished by the trailing flow which follows behind the main discharge.

Water closet trap and discharge pipework does not run full bore, therefore self-siphonage is unlikely.

For further information see BRE Digests 248 and 249.

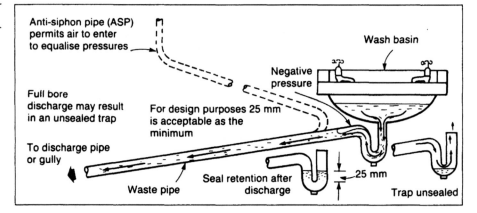

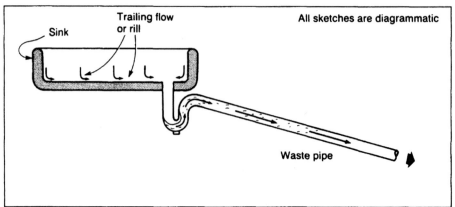

INDUCED SIPHONAGE

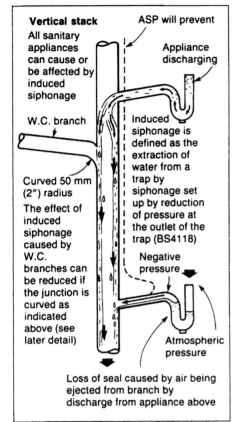

Vertical stack

All sanitary appliances can cause or be affected by induced siphonage

ASP will prevent

Appliance discharging

W.C. branch

Curved 50 mm (2") radius

The effect of induced siphonage caused by W.C. branches can be reduced if the junction is curved as indicated above (see later detail)

Induced siphonage is defined as the extraction of water from a trap by siphonage set up by reduction of pressure at the outlet of the trap (BS4118)

Negative pressure

Atmospheric pressure

Loss of seal caused by air being ejected from branch by discharge from appliance above

Example of loss of seals by induced siphonage

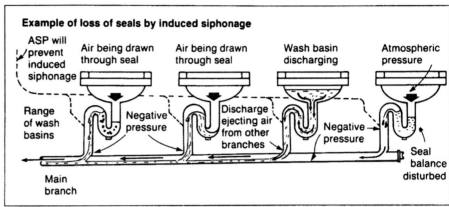

ASP will prevent induced siphonage

Air being drawn through seal

Air being drawn through seal

Wash basin discharging

Atmospheric pressure

Range of wash basins

Negative pressure

Discharge ejecting air from other branches

Negative pressure

Seal balance disturbed

Main branch

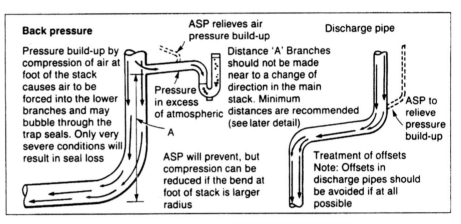

Back pressure

Pressure build-up by compression of air at foot of the stack causes air to be forced into the lower branches and may bubble through the trap seals. Only very severe conditions will result in seal loss

ASP relieves air pressure build-up

Discharge pipe

Pressure in excess of atmospheric

Distance 'A' Branches should not be made near to a change of direction in the main stack. Minimum distances are recommended (see later detail)

A

ASP will prevent, but compression can be reduced if the bend at foot of stack is larger radius

ASP to relieve pressure build-up

Treatment of offsets
Note: Offsets in discharge pipes should be avoided if at all possible

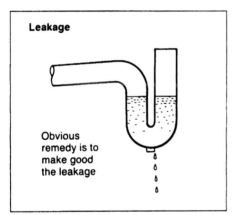

Leakage

Obvious remedy is to make good the leakage

Methods of Prevention

(1) Good design and workmanship as for example 'single stack' plug partial additional venting for stacks in tall buildings.
(2) Fully venting or venting where considered necessary.
(3) Use of re-sealing traps for waste appliances.
(4) Use of larger diameter pipe/s (in certain cases).

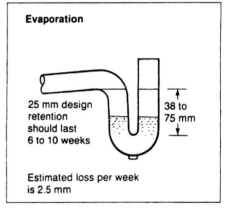

Evaporation

25 mm design retention should last 6 to 10 weeks

38 to 75 mm

Estimated loss per week is 2.5 mm

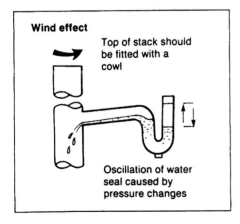

Wind effect

Top of stack should be fitted with a cowl

Oscillation of water seal caused by pressure changes

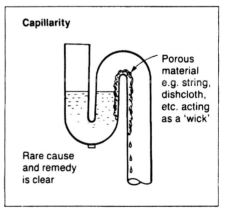

Capillarity

Porous material e.g. string, dishcloth, etc. acting as a 'wick'

Rare cause and remedy is clear

The Two-Pipe System

DEFINITIONS

This system is now considered obsolete, but it has been included in this book as a historical reference.

A soil and waste system comprising two independent pipes, namely, a soil pipe conveying soil directly to the drain and a waste pipe conveying waste water to the drain through a trapped gully*. The system may also require ventilating pipes. Soil is defined as the discharge from water closets, urinals, slop hoppers, stable yard or cowshed gullies and similar appliances. Waste is defined as the effluent from a waste appliance. The latter is defined as a sanitary appliance for the reception of water for ablutionary cleansing or culinary purposes and its discharge after use, e.g. wash basins, sinks, baths, bidets etc. (BS4118).

Notes on the Two-Pipe System. This is the traditional system used in the U.K. and has always been regarded as being simple yet certain, although costly. It is suitable for buildings (all types) which have sanitary appliances widely separated and close grouping together around one mainstack is impracticable. Even at the design stage, this system can become complicated due to the number of pipes involved, viz: soil pipe, waste pipe, often two main ventilating pipes and sundry branch soil, waste and anti-siphon pipes. Before the introduction of Building Regulations, hopper heads could be used in certain positions (if permitted by L.A.) and certain branch waste pipes could then simply discharge into a hopper head. As Reg. N5 requires soil and waste pipes to be internal, hopper heads are no longer permitted for waste drainage. Due to internal fouling, hopper heads can be a source of nuisance and their elimination is welcomed.

LAYOUT EXAMPLES

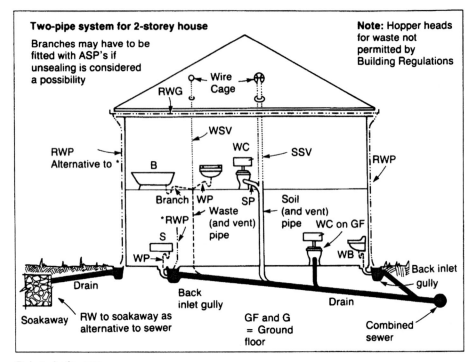

Two-pipe system for 2-storey house

Branches may have to be fitted with ASP's if unsealing is considered a possibility

Note: Hopper heads for waste not permitted by Building Regulations

Points in favour
1. Close grouping of appliances is not essential.
2. Flow considerations are simplified.
3. Simple design and installation with some flexibility for the connection of additional appliances at later stage.

Points against
1. More expensive than other systems.
2. Hopper heads (if permitted and installed) could have been a source of nuisance due to unpleasant odours emanating from the interior of head and pipework.
3. Gullies may also be a source of nuisance due to splashing; blockage; unpleasant appearance or by giving off objectionable odours.
4. Due to the amount of pipework involved, maintenance costs can be high.

	Depth of Seal	
I.D. of trap	Two-pipe	One-pipe
32-64 mm	38 mm	75 mm
75-100 mm	50 mm	50 mm
Minimum seal depths		

Changes in design
1. 'Hopper' (rainwater) heads receiving waste discharges no longer permitted.
2. Optional to discharge waste pipes to drain direct or via trapped gully.
3. No pipework to be placed outside external walls except vent pipes and waste pipes from GF appliances to gullies.
4. Puff pipes should be regarded as obsolete.

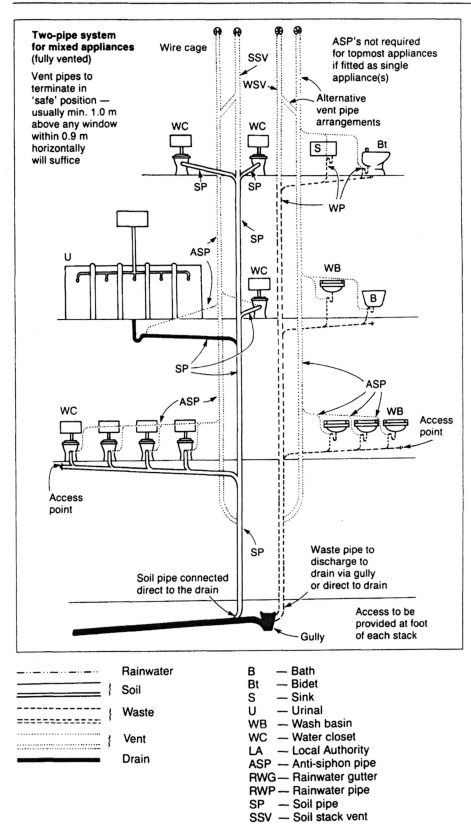

Two-pipe system for mixed appliances (fully vented)

Vent pipes to terminate in 'safe' position — usually min. 1.0 m above any window within 0.9 m horizontally will suffice

Wire cage

SSV

WSV

ASP's not required for topmost appliances if fitted as single appliance(s)

Alternative vent pipe arrangements

WC

WC

S

Bt

SP

SP

WP

SP

U

ASP

WC

WB

B

SP

ASP

WC

ASP

WB

Access point

Access point

SP

Soil pipe connected direct to the drain

Gully

Waste pipe to discharge to drain via gully or direct to drain

Access to be provided at foot of each stack

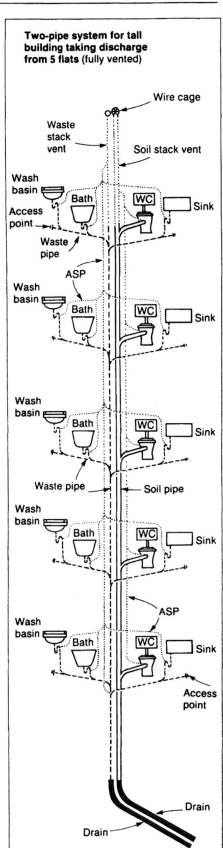

Two-pipe system for tall building taking discharge from 5 flats (fully vented)

Wire cage

Waste stack vent

Soil stack vent

Wash basin

Bath

WC

Sink

Access point

Waste pipe

ASP

Wash basin

Bath

WC

Sink

Wash basin

Bath

WC

Sink

Waste pipe

Soil pipe

Wash basin

Bath

WC

Sink

ASP

Wash basin

Bath

WC

Sink

Access point

Drain

Drain

— · — · — Rainwater	B — Bath
════════ } Soil	Bt — Bidet
- - - - - - } Waste	S — Sink
-=-=-=-=-	U — Urinal
·············· } Vent	WB — Wash basin
▬▬▬▬ Drain	WC — Water closet
	LA — Local Authority
	ASP — Anti-siphon pipe
	RWG — Rainwater gutter
	RWP — Rainwater pipe
	SP — Soil pipe
	SSV — Soil stack vent
	WP — Waste pipe
	WSV — Waste stack vent

Single Stack System

DEFINITIONS

'A soil and waste system in which a single soil-waste pipe conveys soil and waste water directly to the drain. The system may also require ventilating pipes' (BS4118)

In this system, sometimes referred to as the combined system, all discharge pipes conveying soil from water closets, urinals, slop hoppers, etc. and waste from wash basins, sinks, baths, bidets, etc. are connected to one common vertical stack which is connected at the foot directly to the drain similarly to the separate soil pipe in the two pipe system. The provision of ventilating pipes is in accordance with the size of the installation. It should be noted that CP304 recommends the use of the term discharge pipe: 'as a comprehensive all-embracing description in place of the traditional soil and waste terms.'

Notes on the Single Stack System. This system was introduced into this country from the USA about 40 years ago. The idea of discharging all sanitary appliances into one main stack, relying solely upon appliance trap seals to act as foul air barrier, did not meet with the approval of many designers and certain byelaws did not even permit the connecting of a waste pipe to a soil pipe. The one-pipe system had certain advantages over the two pipe system but the risk of trap seal loss had to be overcome. This was achieved by insisting upon deeper seal traps (where possible) and fully ventilating the systems. Eventually, byelaws were amended to permit the use of the one-pipe system. To be an economical proposition at all, close grouping of appliances around the main stack is essential. Therefore, the one-pipe system is suitable for multi-storey dwellings, office buildings, etc. where close grouping is practicable and often repeated at various floor levels. Economy is possible with the larger installation but for 2 storey dwellings, the two-pipe system was less costly when using hopper heads and external downpipes to gullies. As soil and waste pipes must be internal the cost of two-pipe systems for housing has increased. Most designers would now prefer the simplicity and economy of a one-pipe system without anti-siphon pipes if possible, referred to as the single stack system.

LAYOUT EXAMPLES

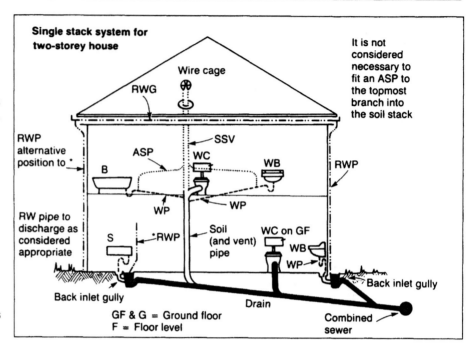

Single stack system for two-storey house

It is not considered necessary to fit an ASP to the topmost branch into the soil stack

GF & G = Ground floor
F = Floor level

In order to ensure that trap seals remain intact whilst imposing no restrictions on positioning appliances, lengths, slope and size of pipes, most branch pipes must be fitted with anti-siphon arrangements.

Internal Diameter of Trap	Depth of seal	
	Two-pipe	One-pipe
32-64 mm	38 mm	75 mm
75-100 mm	50 mm	50 mm

Minimum seal depths

Points in favour
1. Open gullies for waste discharge can be eliminated except for surface water drainage.
2. The elimination of certain pipework i.e. one main stack instead of two.
3. Economy in certain buildings but not necessarily for low rise housing.
4. Simple to design as an internal system.
5. Improvement of appearance when fitted externally (an obsolete advantage).
6. Use of hot water discharges and larger quantities of water reduce deposits.
7. Reduction in maintenance costs.

Points against
1. Sanitary appliances must be grouped around the main stack to be economical.
2. Does not permit the flexibility in design enjoyed by the two-pipe system.
3. As the only barrier to foul air is the appliance trap seal, the importance of efficient trapping and ventilation must be realised.
4. Must use deep seal traps and fully ventilate the system to be 'certain'.
5. High standard design and workmanship required.
6. A blockage at the foot of the main stack will affect the whole system.

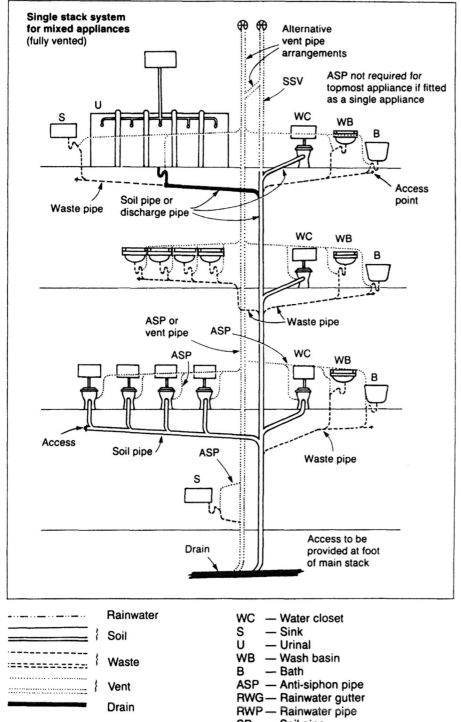

Single stack system for mixed appliances (fully vented)

Alternative vent pipe arrangements

SSV

ASP not required for topmost appliance if fitted as a single appliance

S U

WC WB B

Access point

Waste pipe

Soil pipe or discharge pipe

WC WB B

ASP or vent pipe

ASP

Waste pipe

ASP

WC WB B

Access

Soil pipe

ASP

S

Waste pipe

Drain

Access to be provided at foot of main stack

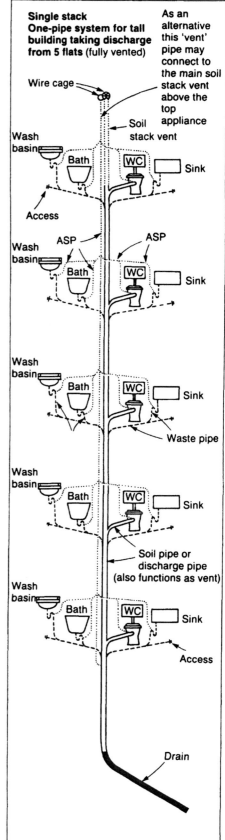

Single stack One-pipe system for tall building taking discharge from 5 flats (fully vented)

As an alternative this 'vent' pipe may connect to the main soil stack vent above the top appliance

Wire cage

Soil stack vent

Wash basin

Bath WC Sink

Access

Wash basin

ASP ASP

Bath WC Sink

Wash basin

Bath WC Sink

Waste pipe

Wash basin

Bath WC Sink

Soil pipe or discharge pipe (also functions as vent)

Wash basin

Bath WC Sink

Access

Drain

Rainwater

Soil

Waste

Vent

Drain

WC — Water closet
S — Sink
U — Urinal
WB — Wash basin
B — Bath
ASP — Anti-siphon pipe
RWG — Rainwater gutter
RWP — Rainwater pipe
SP — Soil pipe
SSV — Soil stack vent

MINIMUM SIZE FOR INDIVIDUAL BRANCHES AND MAIN STACKS

Appliance or pipe Sanitary appliance	Building Regulation. Approved Document H1/BS.5572	BS.5572 Minimum trap sizes	Remarks
Wash basin	32 mm minimum	32 mm	32 mm will suffice in most circumstances
Bath	40 mm	40 mm	Sizes vary: 32, 38, 50 mm (1¼", 1½", 2"). 38 mm is generally satisfactory except for public baths, when 50 mm should be used
Sink	40 mm	40 mm	
Wash tub (domestic)	No mention	50 mm	Normally 50 mm will suffice
Bidet	32 mm	32 mm	Normally 32 mm will suffice
Shower tray	40 mm	40 mm	
Drinking fountain	No mention	32 mm	
Note: Minimum gradient recommended is 1 in 48 (1¼ deg.; 21 mm per metre)			
Water closet	75 mm minimum	Integral trap	BS washdown pan: 88 mm siphonic WC pans: 75 or 82 mm outlet size
Ventilating pipes			
Discharge pipe (for appliances other than urinals)	75 mm	75 mm	Minimum size is normally 88 mm to correspond with outgo size of WC pan
Discharge pipe (for one or more urinals) bowl type	50 mm (2")		Up to 2 urinals: 50 mm is sufficient. 3 to 7 urinals: 65 mm. Over 7 urinals, 2 outlets should be used. (In hard water districts, these sizes should be one size larger)
slab type	65 mm	65 mm	
Stack pipe	75 mm minimum	See BS.5572 75 mm minimum	
Note: Public Health Act 1936 Section 40(2) requires the soil (discharge) pipe from every WC to be properly vented and this is usually done by using a vent pipe of similar size to the soil stack comprising the lower portion of the stack.			

BRANCH VENTILATION PIPES AND STACKS

BS.5772 recommends that the size of ventilating pipes to branches from single appliances can be 25 mm diameter (up to 15m lengths or maximum of five bends). If a greater length or the number of bends exceeds five, a pipe diameter of 32 mm should be used. Where ventilating pipes are at risk, a blockage on a WC branch by continual splashing or submergence, the diameter should be larger. This size can be reduced when above the spill over level of the appliance.

Termination of stack pipe. Should terminate above the level of the eaves or flat roof and not less than 0.9m above the head of any window within a horizontal distance of 3m from the upper end of the vent pipe.

Data Concerning Water

		One litre equals:-
1 pint	= 0.568 l	
1 quart	= 1.137 l	1 dm³, 1000 grammes (ORM)
1 gallon	= 4.546 l	1 kg, 2.205lb.
1ft³	= 28.32 l	0.001 m³, 61.02m³
1yd³	= 0.765m³	0.035 ft³, 1.76 pints
1 m³	= 1000 l	0.22 gal, 1,000,000 mm³

Pressure 1 bar = 1000 mbar or 10⁵ N/m², or 100 kN/m² or 14.5 lbf/in² or 33.45 ft 'head' or 10.2 m 'head'.
1 lbf/in² = 6894.76 N/m² or 6.895 kN/m² or 2.3ft 'head' or 68.95 mbar.
Flow rate 1 gal/min = 0.07577 l/s or 0.272 m³/h. 1 gal/person per day = 4.546 lpd
1 cubic foot/second (cusec) = 28.32 dm³/s (l/s)
1 cubic foot/minute (cumin) = 0.03m³/minute

Contents of bath, cylinders or cistern	
gal.	litre
5	22.7
10	45.4
15	68.2
20	91
25	114
30	136
35	159
40	182
45	205
50	227
60	273
80	364
100	455

Sanitation: Sanitary Pipework Above Ground

VENTILATING STACK SIZES (mm) For commonly used arrangements of discharge stacks and swept entry bends.

Discharge stack size	75 mm	100 mm															150 mm													
Frequency of use	20 min	20 min	10 min								5 min						20 min		10 min		5 min									
Usage description	Domestic	Domestic	Hall of residence				Commercial				Congested						Domestic		Hall of residence		Commercial				Congested					
Number of floors	1 to 2	1 to 3	1 to 10		11 to 15		1 to 8		9 to 12		1 to 4		5 to 8		9 to 12		1 to 30		1 to 30		1 to 8		9 to 24		1 to 8		9 to 16		17 to 24	
Arrangements (see figure 2)	Aa Ab Ba Bb	Aa Ab Ac Ba Bb	Ca	Cb	Ca	Cb	Da	Db	Da	Db	Ea	Eb Ec	Ea	Eb Ec	Ea	Eb Ec	Ca	Cb	Da	Db	Ea	Eb Ec	Ea	Eb Ec	Ea	Eb Ec	Ea	Eb Ec	Ea	Eb Ec
Number of appliance groups (per floor) — 1	0	0	0	32*	50	50	0	32*	32	32	0	32*	0	32*	32	32	0	32*	0	32*	0	32*	0	32*	0	32*	0	32*	0	32*
2		0	0	32*	50	50	0	32*	32	32	0	32*	0	32*	50	50	0	32*	0	32*	0	32*	0	32*	0	32*	0	32*	0	32*
3											0	32*	32	32	40	40					0	32*	0	32*	0	32*	0	32*	65	65
4									0	32*	40	40	40	40	32	32*					0	32*	75	75*	0	32	75	75	75	75
5									0	32*	40	40			32	32*					0	32*	75	75	0	32	75	75	75	75

*Modified single stack arrangement

Connections from the ventilating stack to the discharge stack required on each floor level except where indicated by *

For non-swept connections WC branch fittings (domestic use only):

100 mm discharge stack, single stack system up to four floors. For appliance layouts A1, A2, B1, B2, C1, C2. 150 mm discharge stack, single stack systems up to 15 floors. Two groups of appliances per floor.

The following are conversions to be used with the above table

WC		Urinal		Wash basin			WC		Wash basin
2	+	1	+	2	Equivalent	2	+	2	
2	+	2	+	3	to:	3	+	3	
3	+	3	+	4		4	+	4	
4	+	4	+	5		5	+	5	

References:

Arrangement
Aa   Single storey bungalow
Ab   Single storey bungalow
Ac(i)   Single storey bungalow ⎫ These require certain
Ac(ii) Single storey bungalow ⎬ precautions on the venting of the underground drainage system
Ba   Two- or three-storey house
Bb   Two- or three-storey house
Ca   Multi-storey flats
Cb   Multi-storey flats
Da   Multi-storey halls of residence
Db   Multi-storey halls of residence
Ea   Commercial and public buildings
Eb   Commercial and public buildings
Ec   Commercial and public buildings
BS.5572: 1978 gives illustrations.

Drains and private sewers Building Regulation minimum sizes	
For conveying soil water	100 mm
For conveying trade effluent	100 mm
For conveying waste water	75 mm
For conveying surface water	75 mm
For conveying subsoil water	No mention

MISCELLANEOUS DATA

Water consumption per single use of sanitary appliance		
Sink bowl	18 litre (4 gal)	WC 4.6 litres (1 gal) per flush
Wash basin	7 litre (1.5 gal)	WC 9.1 litres (2 gal) per flush
Bidet (approx.)	5 litre (1.0 gal)	WC 13.6 litres (3 gal) per flush
Bath (depends upon user)		Shower (flow rate): 0.11 l/sec. (1.5 gal/min)

SLOPE (or gradient, inclination or fall)

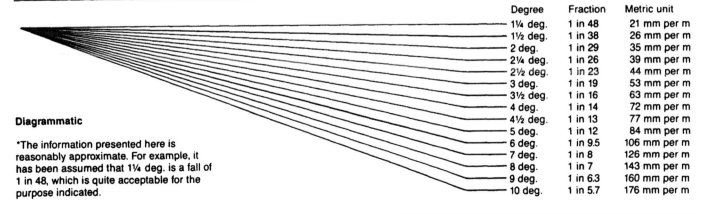

Degree	Fraction	Metric unit
1¼ deg.	1 in 48	21 mm per m
1½ deg.	1 in 38	26 mm per m
2 deg.	1 in 29	35 mm per m
2¼ deg.	1 in 26	39 mm per m
2½ deg.	1 in 23	44 mm per m
3 deg.	1 in 19	53 mm per m
3½ deg.	1 in 16	63 mm per m
4 deg.	1 in 14	72 mm per m
4½ deg.	1 in 13	77 mm per m
5 deg.	1 in 12	84 mm per m
6 deg.	1 in 9.5	106 mm per m
7 deg.	1 in 8	126 mm per m
8 deg.	1 in 7	143 mm per m
9 deg.	1 in 6.3	160 mm per m
10 deg.	1 in 5.7	176 mm per m

Diagrammatic

*The information presented here is reasonably approximate. For example, it has been assumed that 1¼ deg. is a fall of 1 in 48, which is quite acceptable for the purpose indicated.

The Single Stack System (1)

DEFINITIONS

A one pipe system from which, subject to the observance of certain stringent rules, all or most of the trap ventilating pipes are omitted (BS.4118). The single stack system was developed by the Building Research Establishment and was formerly known as the 'simplified system'. The research, which commenced towards the end of the war, soon proved that the unsealing of traps did not occur as readily as had been presumed previously and that, under certain circumstances, trap vents could be omitted from the one pipe system. Properly designed it becomes basically a one-pipe system without vent pipes, hence single stack. Venting is necessary if two branches combine to form common branch.

SINGLE STACK SYSTEM

Design of individual branches
Originally, this system was developed mainly for low cost housing, but having proved successful, it was further developed for multi-storey buildings. The design of the pipework is very important and its success relies upon close grouping of single appliances (each with separate discharge branch) around the stack. Providing the design is correct, vent pipes can be omitted except venting via the mainstack, which continues upwards to a safe position. It produces the simplest possible system for housing as well as high rise domestic and public buildings.

BS.5572:1978 provides information on basic systems in present use, which includes the single stack system as well as a 'modified' single stack system. The Building Research Establishment, which is credited with the early development of the single stack system, produced a two-part document in April 1981 entitled Sanitary Pipework: Part 1 — Design Basis; and Part 2 — Design of Pipework. This work incorporated the basic performance criteria contained with BS.5572 as well as standardising the terminology for describing the parts of the above ground drainage systems. It retains the term 'Simplified (i.e. unvented) System' of drainage and supersedes BRE Digests 80 and 115.

BASIC PRINCIPLES

As explained in earlier detail sheets, the effects of self and induced siphonage can be controlled by the correct choice of pipe size, gradient and bends as well as

DESIGN OF BRANCH DISCHARGE PIPES

the limitation of numbers of appliances connected to the stack. It has been found that if pressure fluctuations within the system are limited to ±375 N/m², the water seal depths of 75 mm for all appliances except the WC (initial seal 50 mm) will remain intact, provided the pipework and fittings are to the appropriate British Standards.

Design curve for nominal 32 mm branch pipe and 75 mm deep seal P-traps connected to single BS wash basins. The minimum desirable slope is 20 mm/m (about 1.25 deg.). The slopes recommended are allowable maxima, pipes need not be fixed at exactly the gradients given on the graph.

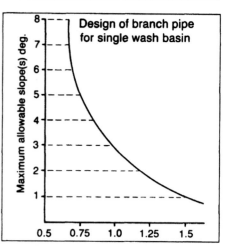

Design of branch pipe for single wash basin

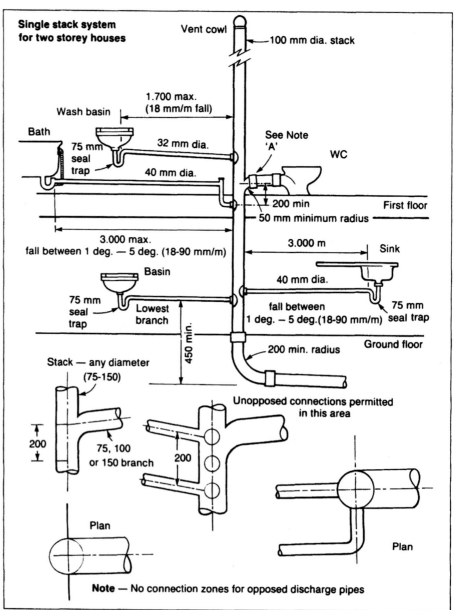

Single stack system for two storey houses

Vent cowl — 100 mm dia. stack

1.700 max. (18 mm/m fall)

Wash basin

Bath

75 mm seal trap — 32 mm dia.

40 mm dia.

See Note 'A'

WC

200 min — First floor

50 mm minimum radius

3.000 max. fall between 1 deg. — 5 deg. (18-90 mm/m)

3.000 m — Sink

Basin

40 mm dia.

75 mm seal trap — Lowest branch

fall between 1 deg. — 5 deg. (18-90 mm/m)

75 mm seal trap

450 min.

200 min. radius — Ground floor

Stack — any diameter (75-150)

200

75, 100 or 150 branch

200

Unopposed connections permitted in this area

Plan

Plan

Note — No connection zones for opposed discharge pipes

BRE Digest 249 provides the following table for the design of branch discharge pipes for both single and multiple appliances.

Appliance	Number of appliances	Trap size mm and type	Branch pipe		Bends in branch pipe	Design precautions			
			Max. length m	Size mm		Principle to be guarded against	Trap design	Branch gradients	Other requirements
Wash basin	Single	32 P	1.7	32	Not more than 2 (min radius 75 mm to centre line)	Self-siphonage	Tubular	See Fig. 6	Recommendations for BS.1188 wash basins with overflow. Some bottle traps will be suitable.
	Single	32 P	3.0	40	Not more than 2 (min radius 75 mm to centre line)	Self-siphonage and excessive deposition	Tubular or bottle	1 deg. to 2½ deg. (18 to 45 mm/m)	Connect 40 mm branch to trap with 32 mm tail pipe. 50 mm long
	Single	32 P or S	3.0	32	Any bends to have min radius of 75 mm to centre line	Self-siphonage and excessive deposition	Resealing	1 deg. to 2½ deg. (18 to 45 mm/m)	See Code of Practice for vented arrangements
	Range of up to 4	32 P	3 Main 0.74 Branch	50 Main 32 Branch	None	Self and induced siphonage and excessive	Tubular or bottle	1 deg. to 2½ deg. (18 to 45 mm/m)	Swept entry fittings at connections between 32 mm and 50 mm branches
	More than 4	32 P	10 Main 1 Branch	50 Main 32 Branch	Any bends to have min radius of 75 mm to centre line	Self and induced siphonage and excessive deposition	Resealing	1 deg. to 2½ deg. (18 to 45 mm/m)	Swept entry fittings at connections between 32 mm and 50 mm branches. See also Code of Practice for vented arrangements
	More than 2	32 S	10 Main 1 Branch	50 Main 32 Branch	Any bends to have min radius of 75 mm to centre line	Self and induced siphonage and excessive deposition	Resealing	1 deg. to 2½ deg. (18 to 45 mm/m)	Swept entry fittings at connections between 32 mm and 50 mm branches. See also Code of Practice for vented arrangements
Bath, Sink	Single	40 P or S	3	40	Any bends to have min radius of 75 mm to centre line	Self-siphonage and excessive deposition	Tubular or bottle	1 deg. to 5 deg. (18 to 90 mm/m)	Longer branches possible but noisier and/or blockage may result. S-trap arrangement will also give noisier discharge
Shower	Single	40 P or S	3	40	Any bends to have min radius of 75 mm to centre line	Self-siphonage and excessive	Tubular or bottle	1 deg. to 5 deg. (18 to 90 mm/m)	
Washing Machine	Single	40 P or running trap	3	40	Any bends to have min radius of 75 mm to centre line	Siphonage from machine during refill	Tubular or bottle	1 deg. to 2½ deg. (18 to 45 mm/m)	40 mm standpipe, 600-900 mm high (depending on machine) with air gap at hose opening
WC	Single	75 to 100 P or S	6	100	Avoid knuckle bends	Excessive deposition	—	1 deg. min (18 mm/m)	Branch pipe size of 75 mm can be used with siphonic WCs
	Range of up to 8	75 to 100 P or S	15	100	Avoid knuckle bends	Excessive deposition and induced siphonage	—	½ deg. to 5deg. (9 to 90 mm/m)	Swept entry fittings needed between WC branches and main branch
Urinal	Single Bowl	40 P or S	3	40	Any bends to have min radius of 75 mm to centre line	Excessive deposition	Tubular or bottle	1 deg. to 5 deg. (18 to 90 mm/m)	Access for cleaning important
	Range of up to 5 bowls	40 P or S	As short as possible	50 Main 40 Branch	Any bends to have min radius of 75 mm to centre line	Excessive deposition and induced siphonage	Tubular or bottle	1 deg. to 5 deg. (18 to 90 mm/m)	Access for cleaning important
	Range of stalls	65 or 75 P or S	As short as possible	65 or 75	Large radius bends	Excessive deposition	Tubular	1 deg. to 5 deg. (18 to 90 mm/m)	Access for cleaning important

The Single Stack System (1)

General rules
1. Pipe runs to be kept as short as possible.
2. Keep pipe gradients uniform.
3. Use of bends to be minimised.
4. Provide access for cleaning blockages.

The frequency of use of sanitary fittings has been taken into account in the formulation of the design tables; as it is considered unreasonable to design a discharge system on the assumption that all or most of the appliances are in use simultaneously. The following table, from BRE Digest 248 shows the assumptions made as well as typical discharge rates for appliances in this country.

Appliance	Capacity litres	Discharge data		Frequency of use (T) s	Individual probability (p) of discharge $p = \frac{t}{T}$
		Maximum flow rate, l/s	Duration (t), s		
Washdown WC	9	2.3	5	1200	0.0041
				600	0.0083
				300	0.0167
Urinal (per person unit)	4.5	0.15	30	1200	0.025
				900	0.0333
Wash basin (32 mm branch)	6	0.6	10	1200	0.0083
				600	0.0167
				300	0.0333
Sink (40 mm branch)	23	0.9	25	1200	0.0208
				600	0.0417
				300	0.0834
Bath (40 mm branch)	80	1.1	75	4500	0.0167
				1800	0.0417
Automatic washing machine	180	0.7	300	15000	0.0200
Shower		0.1			
Spray tap basin		0.06			

Sanitation: Sanitary Pipework Above Ground

DESIGN OF DISCHARGE STACK

Summary of requirements for discharge stacks.

The main factors which can cause pressure fluctuations in vertical pipework are as follows:
(1) Number of appliances and distribution
(2) Pattern of use
(3) Height of building
(4) Dimensions of branch and stack pipes
(5) Design of connections

The following table shows factors to be considered in the design of a discharge stack.

Component	Action to be guarded against	Recommendations
WC (100 mm) branch connections	Induced siphonage	To be swept in direction of flow with minimum sweep radius of 50 mm (but see Note in Table 4)
Bend at base of stack	Back pressure and build up of detergent foam	Bend to be of large radius, at least 200 mm radius to centre line but preferably two 45° large radius bends. Bend can also be enlarged by one size but this may oversize the drain.
Lowest stack connection	Back pressure effects	Distance between lowest branch connection and invert of drain to be: 450 mm for single houses up to three storeys; 750 mm for multi-storey systems up to five storeys. One storey height for systems over five storeys.
Offsets	Pressure effects	Avoid offsets in wet part of stack.
Opposed branch connections	Backing up of WC branch discharge into opposed branch	Position of opposed connection to be as shown in Figure 8

GENERAL COMMENTS

Seal losses produced by flow down the stack depend on the following design considerations — (a) Flow load (depends upon the number of appliances connected to the stack and the frequency with which they are used), (b) diameter of the stack, (c) height of the stack. Excessive seal losses can be prevented by choosing the size of the stack appropriate to the height of the building and to the number of appliances connected to it. Where the layout of appliances is suitable, careful design and installation can lead to considerable economies in pipework, the need for separate vent pipes being eliminated. Where these recommendations cannot be followed, traps should be ventilated by pipes of adequate size.

Some design recommendations
1. All appliances should be individually connected to the main stack which should be at least 100 mm diameter, except for two-storey housing, where 75 mm may be satisfactory.
2. Where sinks are connected to a separate stack, the stack should be larger than normal (minimum 89 mm) for buildings over 5 storeys, and connected directly to a drain.
3. For flats of more than 5 storeys high, ground floor appliances should be connected separately to the drain.
4. Bends and offsets in vertical 'wet' stacks should be avoided.
5. Consider also — (i) materials; jointing, support and protection; (ii) access to pipework, especially joints; (iii) insulation against noise; (iv) risk of freezing; (v) wind effect over the top of stack.
6. Special considerations: (i) relative costs; (ii) space taken up by the pipe(work); (iii) the ease with which the pipes can be accommodated.

Bend at foot of stack

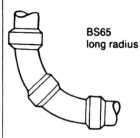

BS65
long radius

To avoid back pressure and build-up of detergent foam, the bend must be of large radius. At least 200 mm radius to the centre line, preferably two 45 deg. large radius bends. A further precaution is to use a bend one size larger than the stack itself.

Wind effects

Suction caused by wind blowing across the tops of stacks on tall buildings can result in excessive pressure fluctuations. Site away from 'edges' of buildings. Also a protective cowl is of some help.

Lowest branch connection

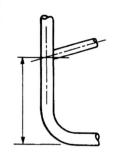

For preference, ground floor appliances should connect directly to the drain. However, for multi-storey systems, distance 'x' should be not less than 0.75 m with 100 mm stack. For 2 storey houses .45 m for a 75 mm stack.

Offsets

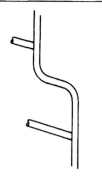

Offsets below the topmost branch connection should be avoided or extra vent pipes may be required to reduce pressure fluctuations in the stack. Offsets in the 'dry' (upper) portion of the stack are acceptable.

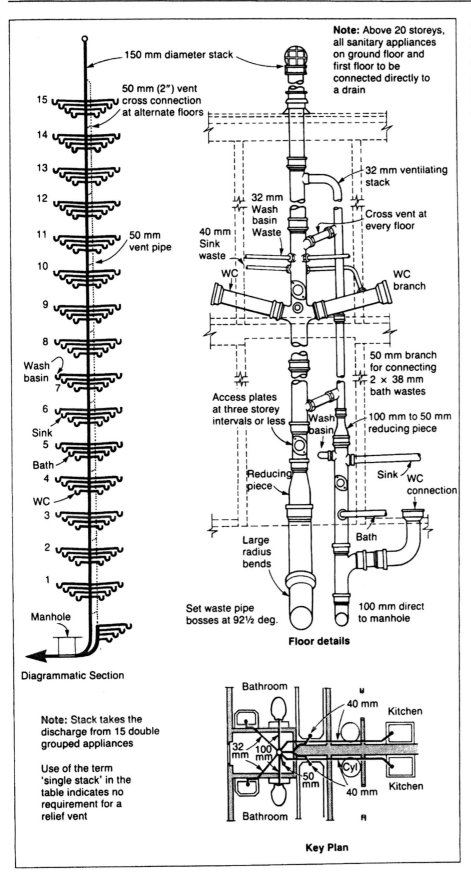

Note: Above 20 storeys, all sanitary appliances on ground floor and first floor to be connected directly to a drain

150 mm diameter stack

50 mm (2") vent cross connection at alternate floors

50 mm vent pipe

15
14
13
12
11
10
9
8

Wash basin
7
6

Sink
5

Bath
4

WC
3
2
1

Manhole

Diagrammatic Section

Note: Stack takes the discharge from 15 double grouped appliances

Use of the term 'single stack' in the table indicates no requirement for a relief vent

32 mm ventilating stack

32 mm Wash basin Waste

40 mm Sink waste

Cross vent at every floor

WC

WC branch

50 mm branch for connecting 2 × 38 mm bath wastes

Access plates at three storey intervals or less

Wash basin

100 mm to 50 mm reducing piece

Reducing piece

Sink

WC connection

Large radius bends

Bath

Set waste pipe bosses at 92½ deg.

100 mm direct to manhole

Floor details

Bathroom

40 mm

Kitchen

32 mm

100 mm

Cyl

50 mm

40 mm

Kitchen

Bathroom

Key Plan

A modified single stack system
Alternatives: (a) 100 mm stack and 75 mm venting or (b) 100 mm stack with 65 mm vent, plus 50 mm cross vent on every floor. (A plumbing system similar in principle has been adopted in 15 storey flats and maisonettes at Comiston Estate, City of Edinburgh). (Design Bulletin 3 Part 1).

Systems for Multi-Storey Office

The terms used in BS.5572 to describe the various parts of above-ground drainage are also used in the BRE Digests 248 and 249 (1981) Sanitary pipework — Part 1 Design basis; Part 2 — Design of pipework. For larger or more complicated systems needing ventilating pipework, reference should be made to the BS.

Arrangement for an eight-storey (maximum) building of commercial use. The layout is suggested by BRE Digest 249.

Maximum number of appliances per floor is five WC's and five washbasins.

Table 2 from BRE Digest 249 refers to BS. 5572 for vent arrangements where more than four basins are required to be connected to stack. Fig. 24(a) (i) of the BS prescribes — maximum length 'L' for discharge branch as 7m; maximum two bends of 75mm minimum diameter (radius to c/1); 25mm ventilating pipe required.

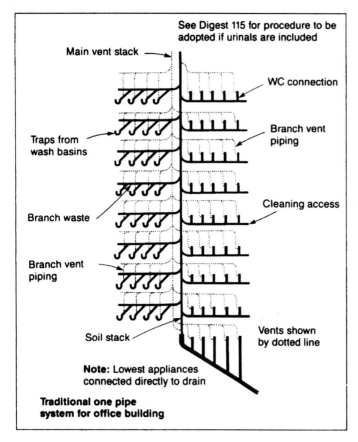

See Digest 115 for procedure to be adopted if urinals are included

Main vent stack

WC connection

Traps from wash basins

Branch vent piping

Branch waste

Cleaning access

Branch vent piping

Soil stack

Vents shown by dotted line

Note: Lowest appliances connected directly to drain

Traditional one pipe system for office building

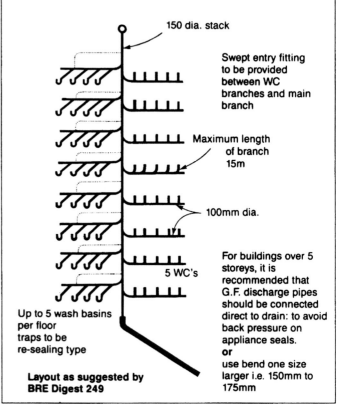

150 dia. stack

Swept entry fitting to be provided between WC branches and main branch

Maximum length of branch 15m

100mm dia.

5 WC's

Up to 5 wash basins per floor traps to be re-sealing type

For buildings over 5 storeys, it is recommended that G.F. discharge pipes should be connected direct to drain: to avoid back pressure on appliance seals. **or** use bend one size larger i.e. 150mm to 175mm

Layout as suggested by BRE Digest 249

Discharge stack sizing
*1 domestic appliance group is 1 WC, 1 sink, 1 wash basin, 1 bath (and/or shower) and 1 washing machine.

Note: For non-swept WC branch connections, with 2 groups of domestic appliances per floor:

A 100mm stack can be used for systems up to 4 floors in height.

A 150mm stack can be used for systems up to 15 floors in height.

The following table gives additional information for offices, and public buildings. The four combinations given are based on the provision to two Statutory Instruments made in the Offices, Shops and Railway Premises Act 1963, and these are equated to the hydraulic equivalents of WC/basin combinations.

Stack size mm	Use		Number of floors of appliances	Max. Number of appliances/floors
	Description	Interval between use mins		
75	Domestic	20	Up to 2	1 Domestic Group*
100	Domestic	20	Up to 10	2 Domestic Groups*
	Commercial	10	Up to 8 Up to 4	2 WCs + 2 wash basins 5 WCs + 5 wash basins
	Congested	5	Up to 8 Up to 4	1 WC + 1 wash basin 3 WCs + 3 wash basins
150	Domestic	20	Up to 30	2 Domestic Groups*
	Commecial	10	Up to 24 Up to 8	3 WCs + 3 wash basins 5 WCs + 5 wash basins
	Congested	5	Up to 24 Up to 16 Up to 8	2 WCs + 2 wash basins 3 WCs + 3 wash basins 5 WCs + 5 wash basins

Sanitation: Sanitary Pipework Above Ground

DESIGN REQUIREMENTS FOR RANGES OF WASH BASINS

WC		Urinal		Wash basin			WC		Wash basin
2	+	1	+	2	Equivalent to:		2	+	2
2	+	2	+	3			3	+	3
3	+	3	+	4			4	+	4
4	+	4	+	5			5	+	5

For situations not covered by the Digests 248 and 249 and British Standards Codes of Practice for sanitary pipework, a test/mock-up is recommended. Test procedures are described in the code of practice.

This appliance is the most likely to risk loss of water seal due to self-siphonage. Single basins fitted with a P-trap can run full bore and so cause self-siphonage. If wash basins discharge pipes exceed 1.7m from the stack, a larger diameter pipe can be used.

Alternatively, the appliance can be fitted with a re-sealing trap, or be vented as described previously.

A range of washbasins can be fitted to a common discharge pipe where spray taps are fitted, up to eight basins have been connected to a 30mm diameter pipe. Access for cleaning should be provided as risk of blockage by sedimentation is possible. Where spray taps are not fitted, ranges of up to four can be connected.

Ranges of WC's do not normally run full so there is no need for venting. Up to eight WC's for straight branch — 100mm diameter at ½ deg. to 5 deg. gradient (9mm to 90mm per m), maximum length of branch pipe — 15m. Swept entry fittings needed between WC branches and main branch.

Urinals. Little risk of self-siphonage, problems are more likely to be caused by sediment in the discharge pipework. Regular cleaning is recommended, particularly in hard water areas. Up to five bowls can be provided where the branch is 40mm diameter and the main branch is 50mm diameter.

Sinks, baths and showers. These appliances are generally free from the effects of self-siphonage as they are usually designed with 40mm branch connections, any reduction of the water seal after use is refilled from water draining from the flat bottom of the appliance.

Washing machines. During discharge it is necessary for the washing machine hose to be placed into a vertical stand pipe (usually 40mm). Connection is normally made to a stack via a running trap. Provided sufficient space is left for air to enter the open end of the stand pipe, siphonage should not occur.

Rainwater connections. It is advised that where the discharge stack is used to carry rainwater (maximum 40m² of roof area), the building height should not exceed ten storeys, and — if accessible — the roof outlet to the stack should be tapped.

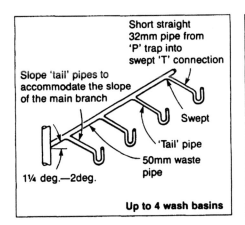

Short straight 32mm pipe from 'P' trap into swept 'T' connection

Slope 'tail' pipes to accommodate the slope of the main branch

Swept

'Tail' pipe

50mm waste pipe

1¼ deg.—2deg.

Up to 4 wash basins

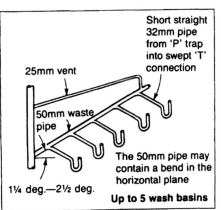

Short straight 32mm pipe from 'P' trap into swept 'T' connection

25mm vent

50mm waste pipe

The 50mm pipe may contain a bend in the horizontal plane

1¼ deg.—2½ deg.

Up to 5 wash basins

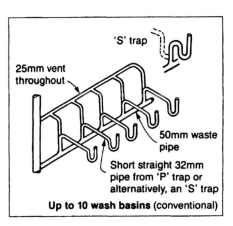

'S' trap

25mm vent throughout

50mm waste pipe

Short straight 32mm pipe from 'P' trap or alternatively, an 'S' trap

Up to 10 wash basins (conventional)

INTRODUCTORY NOTES

Comprehensive design information has been published in several Building Research Establishment Digests and the BS.5572:1978 "Code of Practice for Sanitary Pipework", on single stack systems for domestic buildings. For other types of buildings, certain information has been limited to data on hydraulic loading and little is said in the present BS.5572 about venting requirements in relation to hydraulic loading for other than domestic buildings.

One accepted method for the determination of above-ground drain pipe sizes is the fixture or discharge unit concept in which numerical values are assigned to sanitary appliances to express their load producing properties. The principle was first put foward by Hunter (USA) in an attempt to overcome the problem of calculating the likely hydraulic load with mixed sanitary appliances.

In a mixed system, the sum of all the likely flows for each group or type of appliance gives an over-estimate of the total flow, mainly because the peak usage periods of each appliance type do not normally coincide. The discharge unit concept partially overcomes this objection by a numerical weighting enabling a flow equivalent from a summation of discharge units to be found.

By fixing a flow limit of one quarter capacity in a vertical stack pipe and half capacity for branch discharge pipes based on recommendations of American research workers, this method has been used in BS.5572 to state the maximum discharge units permissible for a given stack size. As implied above, the disadvantage of such a method is that stack sizes are based entirely on hydraulic loadings and on the pressures developed within the stack; it does not give guidance on venting requirements. Tables developed by Wyly & Eaton use some information on the quantity of air drawn down drainage stacks, but they assumed that the vent stack and not the drainage stack carried most of the air in the system. Experience in the UK shows that this assumption is unwarranted and leads to an oversized venting system (Wise & Lillywhite). Other details also deal with stack sizing based on the latest Building Research Establishment theories.

BS.5572 PIPE CAPACITIES

Discharge unit values for sanitary appliances in common use are given in Table 1. For other appliances, the discharge unit value should be taken as that given in the table for an appliance with the same diameter trap with a comparable use interval. Where other intervals are expected, the approximate discharge unit value can be determined since the values given show that the discharge unit value is inversely proportional to the use interval, i.e. if the interval is doubled, then the discharge unit value is halved.

NOTES

(1) The capacity of a vertical discharge pipe (stack) is limited by the need to preserve a large air core to prevent excessive pressure fluctuations. The flow capacity of a stack may therefore be less than that of a pipe of the same diameter laid at a steep fall.

(2) Discharge pipes sized by the BS.5572 method give the minimum size necessary to carry the expected flow load. Separate ventilating pipes may be required. It may be worthwhile to consider oversizing the discharge pipes to reduce the ventilating pipework required.

Sanitation: Sanitary Pipework Above Ground

Table 1 — Discharge Unit Values and Flow Rates for Common Appliances

Type of appliance	Frequency of Use			Discharge units
	Peak domestic	Peak Commercial	Congested	
WC (9 litres)	20			7
		10		14
			5	28
Wash basin	20			1
		10		3
			5	6
Spray tap basin	Allow 0.06 litres/sec per tap			—
Bath	75			7
		30		18
			30	18
Shower	Add 0.1 litre/sec per spray			—
Washing mach. (automatic)	250			4
Sink	20			6
		10		14
			5	27
Urinal (per person)		20*		0.3
			20*	0.3
One Group (1 bath, 1 WC 1 or 2 basins and 1 sink)				14

This method can be used for special installations — e.g. systems for very tall or large buildings not covered by data contained in previous Detail Sheets.

*Frequency of flushing of automatic flushing cistern

Table 2 — Maximum capacity and number of discharge units for vertical stacks (Table 8 from BS.5572)

Size	Approximate capacity of stack (litres/sec)	Approximate number of discharge units
50	1.2	10*
65	2.1	60*
75	3.4	200
90	5.3	350
100	7.2	750
125	13.3	2500
150	22.7	5500

*No WC's

+Not more than one siphonic WC with a 75 mm outlet

Table 3 — Maximum number of discharge units allowed on branch discharge pipes (Table 9 from BS.5572)

Size	Discharge units		
	Gradients		
	½ deg. (9 mm/m)	1¼ deg. (22 mm/m)	2½ deg. (45 mm/m)
32	—	1	1
40	—	2	8
50	—	10	26
65	—	35	95
75	—	100	230
90	120	230	460
100	230	430	1050
125	780	1500	3000
150	2000	3500	7500

Note: Discharge pipes sized in this way give the minimum size necessary to carry the expected flow load. Ventilating pipes may be required, and the following table (Table 10 from BS.5572) gives a general guide for the sizing of ventilating pipes and stacks.

Table 4

Size of branch discharge pipe or discharge stack D	Size of branch ventilating pipe or stack
Smaller than 75 mm	⅔ D (25 mm minimum)
75 mm and above	½ D

INTRODUCTORY NOTES

It must be pointed out that the fixture or discharge unit concept, because it only gives total volume of flow and does not say anything about combinations of actual discharges, the method is inadequate for the new design procedure produced by the Building Research Establishment, which will be dealt with on later Detail Sheets. For mixed appliances, this means an over-estimation of the total likely number of simultaneous discharges and consequently the total flow. This has not been found to be significant, as the appliance producing by far the greatest suction is the water closet; the effect of the remaining appliances is normally quite small. In principle this gives a safety margin. Also certain tall installations are outside the range of the new design procedure and the tables given in BRE Digest 248 can be referred to.

This method is referred to in BS.5572: 1978 as being useful for the calculation of pipe sizes for special installations such as very tall or large buildings which fall outside the scope of normal arrangements. A criterion of satisfactory service of 99.5 per cent is assumed, as explained in the previous Detail Sheet. Numerical values are allocated to specific sanitary appliances, by aggregating the totals of units permissible for a given pipe or stack diameter — a value can be given. Table 4 on the previous Detail Sheet gives a general guide on sizes for ventilating pipe and stack sizes which should be used with this method. It is said that this is a 'safe assumption' and can lead to oversized ventilating systems in some cases.

Total number of discharge units is 202.2. Table 2 on the previous Detail Sheet gives a minimum stack size of 75mm diameter for the approximate number of discharge units. This allows one siphonic WC with a 75mm outlet. However, the stack should not be less than the diameter of the appliances outlets — i.e. 89mm. Therefore, a minimum stack size of 90mm should be used. From this table, it is possible to calculate the size of the ventilating pipes — i.e. a 90mm diameter stack would require a branch ventilating pipe or stack to be 0.5D = 45mm diameter. The nearest commercial size of pipe is 50mm.

Sanitation: Sanitary Pipework Above Ground

WORKED EXAMPLE

Example of pipe sizing by the discharge unit method* on a traditional fully ventilated system for commercial and public buildings

Note: BS.5572 recommends that the minimum size of the main stack should be at the least:- 100mm, with certain exceptions.

Guard must not restrict air flow

0.9 min.

Where less than 3m

Stack can be reduced to 75mm diameter

50mm diameter ventilating stack

Bath, washbasin and sink

50mm diameter ventilating branch

Access

7

Size to be not less than outlet of appliance e.g. 75mm

7   8

14

21

98   84   70   56   42   28   14

50mm diameter   Access

Range of four washbasins
Discharge unit — 3

32mm diameter

3   6   9

50mm diameter main branch

12

145

Range of four washdown WC's
Discharge unit — 14 per WC

No part to be less than outlet size of appliance i.e. 102mm o.d.

'P' or 'S' traps can be used

146.2   1.2   65mm

Access

14   28   42

56   Access   Stub stack

Access

202.2

Could be laid to falls of ½ deg. to 5 deg.

Min. 750mm up to 5 storeys

For commercial and public buildings

Rad.

Minimum 200mm

112

133

90mm minimum diameter stack

Siphonic WC

Fifth floor

Range of eight washdown WC's
Discharge unit — 14 per WC

Outlet size of WC to BS.5503 —
102mm outside diameter

15m max.

Fourth floor

Access

Third floor

Range of 4 urinal stalls
Discharge unit — 0.3 per stall

Second floor

First floor

Ground floor

All ground floor appliances connect to drain direct via stub stack

*Based on information in Burberry P and Griffiths T.J. Demand and discharge pipe sizing for sanitary fittings

Inspection and Testing

GENERAL

Should preferably be done during installation, particularly work which will be concealed. Pre-assembled units can be tested where made and inspected on delivery and/or before fixing. Final tests should be applied on completion of the installation both for soundness and performance. Air or smoke test should be used generally, but, if water be used, then the 'test' should be up to the level of the lowest sanitary appliance. The water test should be used for new systems only.

With old systems incorporating shallow seals, the pressure must be limited accordingly. Any defects revealed should be made good and tested again, until acceptable. Water seal also serves to indicate whether seal is effective or not.

TESTS FOR SOUNDNESS

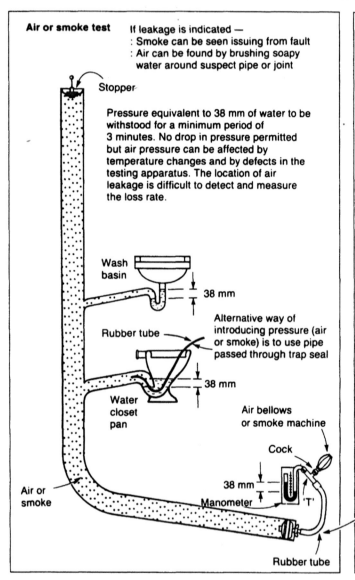

Air or smoke test

If leakage is indicated —
: Smoke can be seen issuing from fault
: Air can be found by brushing soapy water around suspect pipe or joint

Stopper

Pressure equivalent to 38 mm of water to be withstood for a minimum period of 3 minutes. No drop in pressure permitted but air pressure can be affected by temperature changes and by defects in the testing apparatus. The location of air leakage is difficult to detect and measure the loss rate.

Wash basin

38 mm

Rubber tube

Alternative way of introducing pressure (air or smoke) is to use pipe passed through trap seal

38 mm

Water closet pan

Air bellows or smoke machine

Cock

38 mm

Manometer 'T'

Air or smoke

Rubber tube

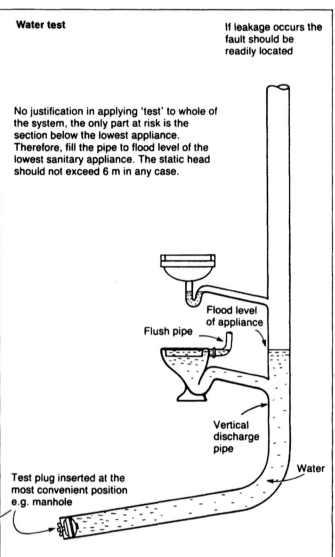

Water test

If leakage occurs the fault should be readily located

No justification in applying 'test' to whole of the system, the only part at risk is the section below the lowest appliance. Therefore, fill the pipe to flood level of the lowest sanitary appliance. The static head should not exceed 6 m in any case.

Flood level of appliance

Flush pipe

Vertical discharge pipe

Water

Test plug inserted at the most convenient position e.g. manhole

Approved Document H1 Section 1.32 — Watertightness states "the installation should be capable of withstanding an air or smoke test of positive pressure of at least 38 mm water gauge for at least 3 minutes. During this time, every trap should maintain a water seal of at least 25 mm.

Sanitation: Sanitary Pipework Above Ground

TESTS FOR PERFORMANCE (BS5572)

Tests are designed to simulate the probable worst conditions in practice with the criterion of 25 mm seal retention when subjected to appropriate discharge test. Repeat each test three times, recharge seal between tests, measure loss with dipstick and take maximum loss of seal as significant result. A reasonable test would be to discharge up to say, 1 W.C. and 1 basin and 1 sink on top floor, with others distributed along the stack, but refer to Tables 1 and 2 for numbers applicable.

Dwellings. To test self-siphonage of waste appliances — fill to overflow, discharge in the usual way and measure seal when finished. This test is important for wash basins but not applicable for W.C. pans which should be flushed in the normal manner. Baths are ignored as their use is spread over a period and consequently do not add materially to the normal peak flow, but a stack that serves baths only, treat same as for sinks.

Public Buildings. Single appliances — test as for domestic. Ranges of appliances — take appropriate combination from Table 2, carry out test and measure seals in all traps in the range. Worst condition is the discharge of appliances at the top end of the range. For ranges in congested use e.g. schools — double the number of appliances taken from table. To test the stability of trap seals when water flows down the stack, discharge selected appliances simultaneously. Measure the remaining seal in the traps when the discharge has ended.

Number of appliances to be discharged simultaneously for testing stability of trap seals

Table 1

Dwellings	Number of appliances to be discharged simultaneously		
Total number of appliances of each kind on the stack	9 litre (2 gal) Water Closet	Wash basin	Kitchen sink
1-9	1	1	1
10—24	1	1	2
25—35	1	2	3
36—50	2	2	3
51—65	2	2	4

Table 2 (This is Table 14 from BS5572. A table for "congested usage" is also provided).

Public Buildings (e.g. office blocks, schools)	Number of appliances to be discharged simultaneously	
Total number of appliances of each kind on the stack	9 litre (2 gal) Water closet	Wash basin
1—9	1	1
10—18	1	2
19—26	2	2
27—52	2	3
53—78	3	4
79—100	3	5

Pipes and fittings should be suitable for their purposes and, where applicable, should comply with the requirements of the relevant British Standards (Clause 6 of BS lists British Standards referred to in document). The choice of materials depends on the size and function of the pipework, the temperature and constituents of the discharge and the ambient conditions, including temperature. Other considerations are the weight, physical strength, ease of assembly and maintenance requirements of the pipework.

Common materials are usually suitable for discharge and ventilating pipes are: cast iron, copper, galvanised steel, lead, and stainless steel. The effects of electrolytic corrosion may occur where dissimilar metals are in contact in the presence of moisture.

Reference should be made to Approved Document H1, Section 1.31 "Materials for pipes, fittings and joints".

The following order of stating the metals commonly used in discharge pipe systems indicates the likely effect of combining any two of them, the former one stated in the 'scale' being the one which is attached — 1. Zinc. 2. Iron. 3. Lead. 4. Brass. 5. Copper. The further apart the metals are in the 'list', the greater will be the attack — e.g. zinc used in the presence of copper results in a severe attack on the zinc.

Notes:
(1) The British Standard for Asbestos Cement soil, waste and vent products has been withdrawn and it may be difficult to find a manufacturer still marketing this material for drainage above ground. However, many A.C. installations are in existence, requiring maintenance as necessary.
(2) PF is not in great demand for this purpose but it is still used by at least one major contractor.

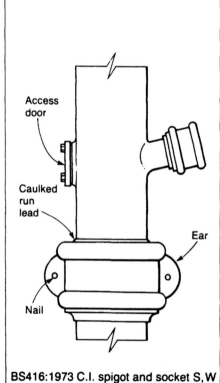

Copper can also be pre-fabricated

Bronze weld

Capillary silver soldered

BS2871:Part 1 1971 Copper tubes for water, gas and sanitation

Access door

Caulked run lead

Ear

Nail

BS416:1973 C.I. spigot and socket S, W and V pipes and fittings

Copper
An adaptable material of medium weight, obtained in 3 grades:- Table X, half hard and jointed by any of a wide range of joints. Table Y, half hard or annealed, jointed as above. Table Z, hard drawn thin wall, not for bending and limited to the use of certain joints. Obtainable in long lengths; expansion is more than cast iron and must be allowed for.

Cast iron (CI)
Has been the most widely used of all types but is heavier than most other materials. Two socket profiles available and pipe can be sand cast or spun. Standard test:- 70kN/m² hydrostatic. Durability depends upon coating of tar based composition or natural bitumens and asphalt with hardener. Wide range of fittings available.

Sanitation: Sanitary Pipework Above Ground

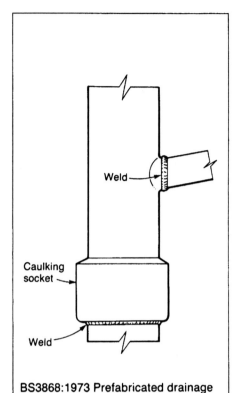

BS3868:1973 Prefabricated drainage
stack units
BS1387:1967 Steel tubes and tubulars

Galvanised steel
Manufactured from light or medium,
black steel tubes and galvanised after
pre-fabrication. Lighter than cast iron.
Smaller sizes easily jointed by screwed
joints. Painting may not be necessary but
is recommended to ensure protection
from atmospheric attack. Protect against
attack from cement, lime, plaster, etc.,
and electrolytic action.

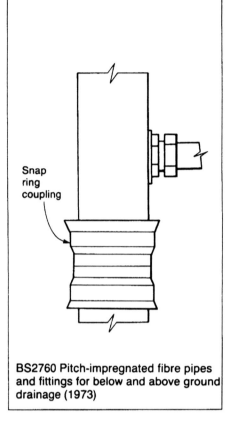

BS2760 Pitch-impregnated fibre pipes
and fittings for below and above ground
drainage (1973)

Pitch fibre (PF)
A light material of preformed, felted
fibrous structure impregnated all through
under vacuum and pressure with coal tar
pitch (e.g.) comprising minimum 65% of
the finished product. Easy to handle, cut
and joint. Should preferably be installed
in ducts of same fire resistance as the
building. Good range of fittings available.
No need to paint.

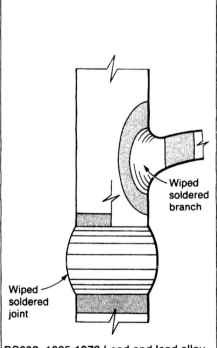

BS602, 1085:1970 Lead and lead alloy
pipes (for other than chemical purposes)
(Another BS is applicable for lead used
for chemical purposes)

Lead
Traditional material for this purpose being
very adaptable, easy to repair but
requires more skill in handling and
jointing than most other materials. Heavy
material, easily damaged and requires
frequent or even continuous support. Use
Table 5 of the BS using chemical
Composition II. Protect against attack by
cement, lime, plaster, magnesite, etc.

Materials, Jointing and Fixing (2)

REFERENCES

Plastics materials. Pipes made of a number of thermoplastics materials are suitable for conveying the discharges from sanitary appliances (also rainwater). Such pipes are termed 'discharge pipes' in the relevant C.P. These include unplasticised polyvinylchloride (uPVC), polypropylene (PP), acrylonitrile butadienestyrene (ABS), polythene — sometimes referred to as polyethylene (PE). All these materials are light in weight and consequently easy to handle and highly resistant to corrosion. Their co-efficients of expansion are, however, much higher than those of metals and proper and adequate allowance must be made for thermal movement. uPVC is the most commonly used plastics material for the larger diameter discharge and ventilating pipes, but should not be used where large volumes of water are discharged at temperatures exceeding 60 deg. C. The short term full bore discharges from some appliances may be at much higher temperatures e.g. some types of washing machines discharge water at 80 deg. C or even higher and the risk of distortion in uPVC pipes is correspondingly greater. Jointing may be accomplished by simple solvent welding, in which case expansion joints may be incorporated in the system. ABS pipes can be used in much the same way as uPVC, but have the advantage of being suitable for conveying discharges at higher temperatures. Polythene pipes, both high and low density, are used mainly for laboratory installations and certain small diameter waste pipes. They are more flexible than uPVC, less liable to impact damage and are resistant to damage by freezing but need adequate support. Jointing is by welding or by mechanical means. Polypropylene pipes can be used in much the same manner as polythene pipes, but they are suitable for conveying discharges at higher temperatures. The correct grade must be used — consult manufacturer(s).

Important Note: Plastics pipes for S.W. and V. pipes have been available commercially for many years and it must be recognised that pipes and fittings will continue to be ordered by 'systems' as offered by various manufacturers. It is not yet possible to provide a standard that would permit complete interchangeability of one manufacturer's products with those of another. In fact, conditions of sale often include ". . . the Company will not be responsible for the malfunctioning of any installation, which includes plastics components not supplied by the Company". Model Specifications are suggested such as "Supply XYZ Ltd. ABS soil and waste pipes and fittings in accordance with the architect's drawings and fix generally according to the manufacturer's instructions".

For details of joint assembly, temperatures, etc., see manufacturer's instructions.

Notes from BS4514 'uPVC Soil and Vent Pipe, Fittings and Accessories'. Material. Substantially of uPVC with additives essential for the manufacture of the polymer. . . to produce sound, durable extrusions or mouldings of good surface finish and mechanical strength and such pigments to provide colour fastness. Black or choice of two approved grey colours. Rubber sealing rings to BS2494 Part 2.

Appendix F of Approved Document B of the Building Regulations 1985 — Protection of Openings — lays down specific requirements for various openings made in fire barriers including pipes, ventilation ducts, and flues. Section F9 states:

METHODS OF JOINTING

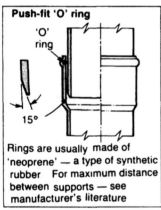

Push-fit 'O' ring

'O' ring

15°

Rings are usually made of 'neoprene' — a type of synthetic rubber For maximum distance between supports — see manufacturer's literature

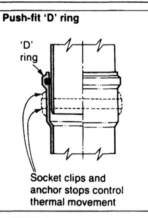

Push-fit 'D' ring

'D' ring

Socket clips and anchor stops control thermal movement

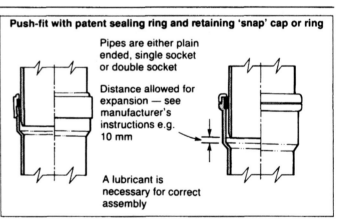

Push-fit with patent sealing ring and retaining 'snap' cap or ring

Pipes are either plain ended, single socket or double socket

Distance allowed for expansion — see manufacturer's instructions e.g. 10 mm

A lubricant is necessary for correct assembly

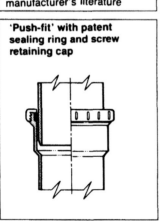

'Push-fit' with patent sealing ring and screw retaining cap

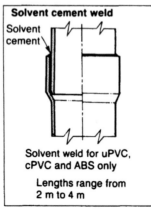

Solvent cement weld

Solvent cement

Solvent weld for uPVC, cPVC and ABS only

Lengths range from 2 m to 4 m

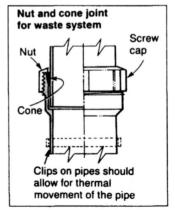

Nut and cone joint for waste system

Nut

Screw cap

Cone

Clips on pipes should allow for thermal movement of the pipe

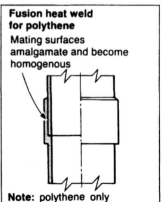

Fusion heat weld for polythene

Mating surfaces amalgamate and become homogenous

Note: polythene only

Sanitation: Sanitary Pipework Above Ground

(a) A proprietary sealing system which has proven ability (by test) to maintain the fire resistance of the wall, floor or cavity barrier.

(b) The nominal internal diameter of the pipe should not exceed the sizes given in Table F1: openings for pipes to be kept as small as possible and the gap between the pipe and wall/floor/cavity barrier to be fire-stopped.

Notes

1. A non-combustible material (such as cast iron or steel which if exposed to a temperature of 800°C will not soften nor fracture to the extent that flame or hot gases will pass through the wall of the pipe.
2. uPVC pipes complying with BS4514:1983.
3. Pipes forming part of an above ground drainage system and enclosed as shown in Diagram F2.

Section F10 states: The diameter of pipes for specification (b) and situation 2 of Table F1 assume that the pipes are part of an above ground drainage system and are enclosed as shown in diagram F2. Where this is not the case, the smaller size diameter referred to in situation 3 should be used instead.

Section F11 states: Where a pipe of specification (b) is used as shown in diagram F3, 'Alternative A', the maximum nominal internal diameter of the sleeving material may be substituted for the diameter of the pipe.

F12 states: Where a pipe comprises sections of different materials, as shown in diagrams F3 'Alternative B' or 'Alternative C', the maximum nominal internal diameter of the pipe may be related to the section of pipe penetrating the wall, floor or cavity barrier.

Table F1 — maximum nominal internal diameter of pipes	Pipe material and maximum nominal internal diameter (mm)		
Situation	Non-combustible material[1] (a)	Lead aluminium or aluminium alloy asbestos-cement or upvc[2] (b)	Any other material (c)
1 Structure (but not a Separating Wall) enclosing a Protected Shaft which is not a stairway or lift shaft	160	110	40
2 Separating Wall between dwelling houses or Compartment Wall or Compartment Floor between flats	160	160 (stack pipe)[3] 110 (branch pipe)[3]	40
3 Any other situation	160	40	40

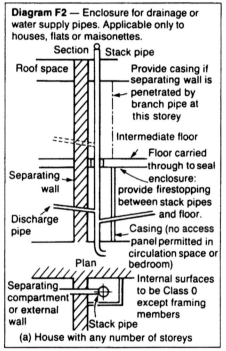

Diagram F2 — Enclosure for drainage or water supply pipes. Applicable only to houses, flats or maisonettes.

Section
Roof space
Stack pipe
Provide casing if separating wall is penetrated by branch pipe at this storey
Intermediate floor
Floor carried through to seal enclosure: provide firestopping between stack pipes and floor.
Separating wall
Discharge pipe
Casing (no access panel permitted in circulation space or bedroom)
Plan
Separating compartment or external wall
Internal surfaces to be Class 0 except framing members
Stack pipe
(a) House with any number of storeys

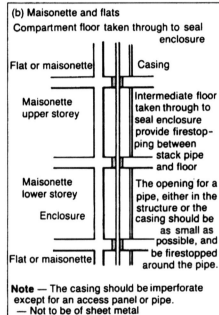

(b) Maisonette and flats
Compartment floor taken through to seal enclosure

Flat or maisonette — Casing
Maisonette upper storey — Intermediate floor taken through to seal enclosure provide firestopping between stack pipe and floor
Maisonette lower storey — The opening for a pipe, either in the structure or the casing should be as small as possible, and be firestopped around the pipe.
Enclosure
Flat or maisonette

Note — The casing should be imperforate except for an access panel or pipe.
— Not to be of sheet metal
— and be ½ hr (min) fire resistance including access door

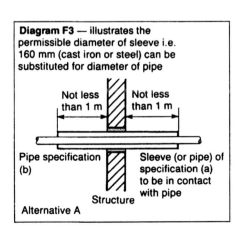

Diagram F3 — illustrates the permissible diameter of sleeve i.e. 160 mm (cast iron or steel) can be substituted for diameter of pipe

Not less than 1 m | Not less than 1 m
Pipe specification (b)
Sleeve (or pipe) of specification (a) to be in contact with pipe
Structure
Alternative A

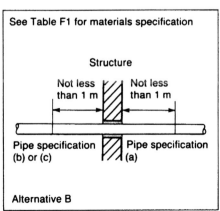

See Table F1 for materials specification

Structure
Not less than 1 m | Not less than 1 m
Pipe specification (b) or (c)
Pipe specification (a)
Alternative B

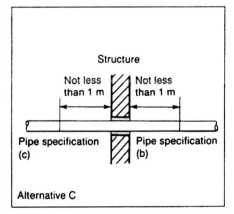

Structure
Not less than 1 m | Not less than 1 m
Pipe specification (c)
Pipe specification (b)
Alternative C

Interpretations
The Building Regulations 1985

Note
Stack pipe: can also convey rainwater
Vent pipe: 'does not convey any soil water, waste water or rainwater'.
Discharge pipe: can also convey rainwater.
Approved Document H3 1.5 of the 1985 Building Regulations states "rainwater pipes should discharge into a drain or gully, but if discharging into a "combined system" — it should do so through a trap. 'No rainwater pipe shall be constructed so as to discharge into or to connect with any pipe or drain used or intended to be used for conveying soil water or waste water, unless provision is made in the design of the sewerage system for the discharge of rainwater'.

An interpretation:- if the sewer into which it is intended to discharge the rainwater has been designed to (and can still) receive rainwater e.g. combined sewer, partially separate sewer; the system can be arranged to receive rainwater, but consult Local Authority).

Table 2 of the Approved Document states that a 100 mm diameter half-round rainwater gutter can drain an effective area of 37 m² with sharp edged outlet of 63 mm diameter.

Size
The capacity of the drainage system should be large enough to carry the expected flow at any point in the system. Table 2 of Approved Document H3 gives typical flow capacities of common size gutters and outlets.

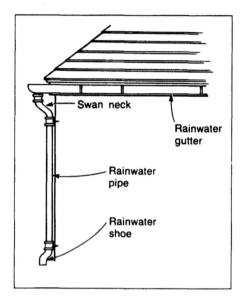

BUILDING REGULATION (Approved Document H3) APPLICABLE TO RAINWATER GUTTERS

Support
Sections 1.7(b) and (c) relate to support. Gutters and rainwater pipes should be firmly supported without restricting thermal movement. Different metals should be separated by non-metallic materials to prevent electrolytic corrosion.

Examples of gutter support brackets
Some variations may exist between different manufacturers products.

Outlets
Section 1.3 gives minimum sizes of outlets for given gutter sizes in Table 2 of the Approved Document. Where the outlet is at the end, the gutter should be sized for the larger areas draining into it, and in this way rainwater pipes may be up to 16 m apart.

Section 1.4 says that gutters should be laid with any fall towards the nearest outlet. If an outlet is rounded, it may be possible to reduce the size of the gutter and pipes. Detailed recommendations can be found in BS6367:1983.

Section 1.6 states that the size of a rainwater pipe should be at least the size of the outlet from the gutter.

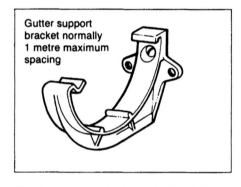

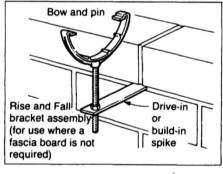

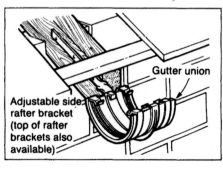

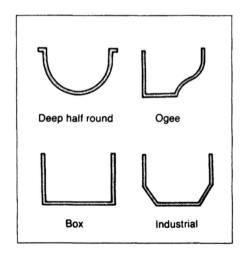

Sanitation: Roof Drainage

Requirements

Materials
Section 1.7 states that any of the materials listed in Table 2 of the Approved Document may be used — for example aluminium, cast iron, copper, galvanised steel, lead, low carbon steel, pressed steel, upvc and zinc.

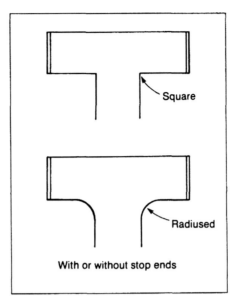

Square

Radiused

With or without stop ends

Joints
Section 1.7(a) states that all gutter joints should remain watertight under normal working conditions.

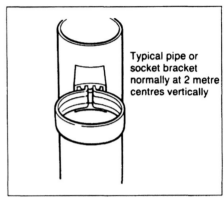

Typical pipe or socket bracket normally at 2 metre centres vertically

APPROVED DOCUMENT H3 APPLICABLE TO RAINWATER PIPES

Outside a building
Size: to be 'of adequate size for its purpose'.

Rule of Thumb: 1 sq. in. of cross-section area of pipe per 100 sq. ft. of 'flat' roof area.*

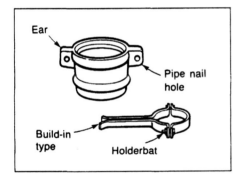

Ear

Pipe nail hole

Build-in type

Holderbat

Support: to be adequately supported throughout its length without restraining thermal movement, any fitting which gives such support being securely attached to the building.
*Metric: 650 mm² per 9 m²

Materials: to be 'of suitable materials of adequate strength and durability'. (A material which is satisfactory for a gutter is usually acceptable for a rainwater pipe).

Examples: Cast iron, asbestos cement, aluminium, uPVC lead, 'steel'

Dampness: to be 'so arranged as not to cause dampness in, or damage to any part of the building'. (Use back inlet gully for preference).

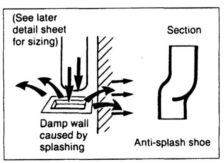

(See later detail sheet for sizing)

Section

Damp wall caused by splashing

Anti-splash shoe

Size: to be of adequate size for its purpose.

Notes from Sanitary Pipework
In areas where the combined system of drainage is permitted it is advantageous to connect roof rainwater outlets directly to discharge pipes providing that:
1. It is practicable and economical to do so and care is taken in positioning the upper terminal point.
2. Normal ventilation of the discharge stack is maintained in case the rainwater outlet should become blocked.
3. No increase in size of discharge stack is necessary, provided the area of roof to be drained does not exceed 40 m²

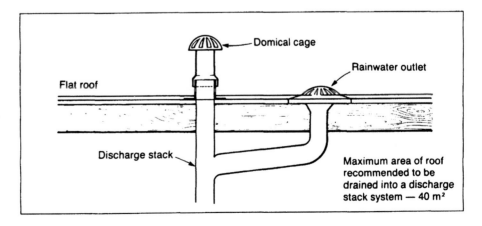

Domical cage

Rainwater outlet

Flat roof

Discharge stack

Maximum area of roof recommended to be drained into a discharge stack system — 40 m²

RELEVANT TERMS FOR ROOF DRAINAGE EXTRACTED FROM BS.4118:1981

Anti-splash shoe
A rainwater fitting fixed at the lower end of a rainwater pipe and so shaped as to reduce splashing when rainwater is discharging into the open air.

Balcony outlet
A fitting intended to be interposed in a vertical rainwater pipe passing through a balcony and providing an inlet for the drainage of rainwater from the balcony.

Cesspool
A box-shaped receiver constructed in a roof or gutter for collection rainwater which then passes into a rainwater pipe connected thereto. (This term also used for 'underground chamber for foul water').

Domical grating
A dome-shaped grating for covering the outlet from a roof (also outlets from urinals and floor channels).

Down pipe
A pipe for conveying rainwater from a roof or other parts of a building (see Rainwater pipe).

Drop end
A gutter outlet, with nozzle and stop-end combined in one fitting (also called stop end outlet).

Fascia bracket
An eaves gutter support designed for screwing to a fascia board.

Gully
A drain fitting or assembly of fittings to receive surface water and/or discharge from waste pipes. It incorporates a trap or sump or both, for the retention of detritus (e.g. grit). The top usually has a grating but may have a sealed cover and can have a back inlet(s) and access to the drain.

Gutter
A channel for collecting rainwater from roofs (colloquial — a channel for collecting surface water).

Boundary wall gutter
A gutter having a flat sole, one upright side and one side angled to suit the slope of an adjacent pitched roof.

Box gutter
A gutter having a flat sole and both sides upright (sole = lower surface or base).

Eaves gutter
A gutter fixed at the eaves (eaves = the lowest, overhanging part of a sloping roof).

Half round gutter
An eaves gutter having a half round cross-section (rhone — see Scottish Building Regulations).

Half round gutter, beaded
A half round gutter with one edge or both edges stiffened by having an integral beading.

Moulded gutter
An eaves gutter having a flat sole, an upright back and an ornamentally shaped front.

Northlight gutter
A valley gutter having a flat sole and sides angled to suit the unequal pitches of a northlight roof.

Ogee gutter
An eaves gutter having an upright back and a combined sole and front of ogee shape.

Parapet gutter
A gutter at the junction of a roof pitch and a parapet wall.

Valley gutter
A gutter having a flat sole and two sides angled to suit the slopes of adjacent pitched roofs (centre gutter).

Hopper head
A flat or angle-backed rainwater head of tapered shape.

Internal spigot joint
A form of eaves gutter jointing which eliminates the projection of external sockets.

Nozzle
A gutter fitting consisting of a short length of gutter in which is formed an outlet tail for connecting to a rainwater pipe (gutter outlet).

Offset
A pipe fitting used to connect two pipes whose axes are parallel but not in line (swanneck).

Pass-over offset
A pipe fitting arranged to permit a pipe to pass over an eaves gutter and follow the pitch of the roof.

Rainwater pipe
A pipe for conveying rainwater from a roof or other parts of a building (downcomer, down pipe, fall pipe, etc.).

Rafter bracket
An eaves gutter support designed for screwing to a rafter (rafter side or top fixing).

Rainwater fitting
A fitting for a rainwater pipe.

Rainwater head
A hopper — or box-shaped rainwater fitting used to collect rainwater for discharging into a rainwater pipe.

Rainwater separator
A device intended, where rainwater is collected from roofs for storage, to divert to waste, the dirty water running off at the beginning of rainfall.

Rainwater shoe
(1) A rainwater fitting at the foot of a rainwater pipe to discharge rainwater into the open air clear of the building surface to which the pipe is fixed. (2) A drain fitting, fixed horizontally at the foot of a rainwater pipe, having a vertical or horizontal inlet and an inspection opening with either a grating or sealed cover.

Roof outlet
A rainwater fitting, normally provided with a grating, for building into a flat roof to receive rainwater for discharging into a rainwater pipe.

Soakaway
A pit dug in permeable ground, filled with hardcore and usually covered with earth, to which liquid is led, and from which may soak into the ground (maybe 'lined' with stone, etc.)

Stop-end
A terminal piece for sealing the closed end of a channel or gutter.

Return stop-end
A stop end for a moulded gutter in which the profile of the gutter is carried round the end of the stop end.

Sanitation: Roof Drainage

Sizing of Rainwater Goods (1)

DESIGN RATE OF RAINFALL

For roof drainage calculations it is usual to assume a rate of rainfall for design purposes of 75mm per hour (1.25mm/min) for the British Isles. Regional differences are not significant except in relation to total rainfall and can be ignored for this purpose. The 75mm/h rate is exceeded on rare occasions and such severe storms are of comparatively short duration with intensities of 150mm/h or more. Where ponding or overflowing cannot be tolerated (except during those very rare storms for which design is impracticable) a design rate of rainfall of up to 150mm may be used. Such increases in the design rates are advisable for buildings where overflow inside would cause serious damage or inconvenience, e.g. museums, power stations and other industrial buildings, particularly if valley gutters are incorporated. An indication of the frequency of various rainfall intensities is as follows:

75mm/hour may occur for:
  5 minutes once in four years
  9 minutes once in ten years
20 minutes once in fifty years

100mm/hour may occur for:
3 minutes once in five years
4½ minutes once in ten years
6½ minutes once in twenty years

Note: 1mm on 1m² = 1 litre/hour
          (÷60 for 1/min)

For intensities other than 75mm/h, substitute 1.25× design intensity (mm/hour) ÷ 75

ROOF AREA ALLOWANCES

The following information is taken from BS.6367: 1983 Code of Practice for drainage of roofs and paved areas.

Effective catchment area for flat roofs and paved areas (Ae)
Where freely exposed horizontal surface is equal to the plan area of the surface. If sloping or vertical areas discharge on to the area, the additional areas should be calculated and added to the plan area.

Sloping roofs
The effective catchment area (Ae) for a freely exposed roof draining to an eaves or parapet wall gutter, is equal to the plan area of the roof plus half its area in elevation.

winds, vertical uplift and other effects. For design purposes, it is assumed that the total rain approaching the wall surface may be used — the resultant figure will normally have a large margin of safety. Where run-off is onto paved surfaces, this can be ignored unless flooding of such areas cannot be permitted. To calculate the effective catchment area, Ae, for vertical surfaces, the following formula may be used.

$$Ae = \frac{1}{2}\sqrt{(A^2v_1 + A^2v_2 - 2Av_1 + Av_2 \cos\theta)}$$

Where Av_1 and Av_2 are areas of the vertical walls as shown contributing to the flow of the gutter.

For a single wall, Ae should be taken as half the exposed area of the wall. Where two or more walls form an angle or bay, the direction of the wind shall be assumed to be acting on the two surfaces equally.

For areas draining into a centre gutter (e.g. 'northlight') it is reasonable to assume that factors concerning pitch can be ignored because excess rain on one slope will be largely offset by less rain on the facing slope. The tables referred to above and the example below are given on Detail 62.

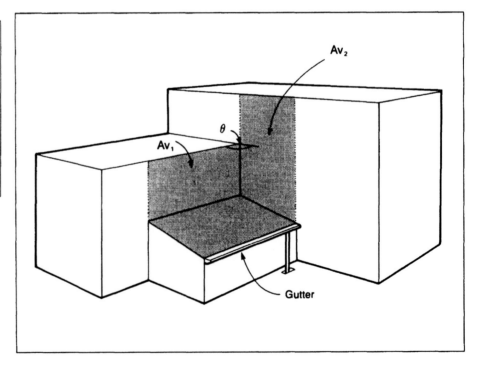

Effective catchment area Ae = Plan area (Ah) + ½ area of elevation (Av)

Note: The Approved Document H3 of the Building Regulations 1985 provides a table for calculation of sloping roofs at varying pitches.

Vertical surfaces
In practical terms, not all wind-driven rain will cause run-off, due to losses from rebound, lateral movement by cross

Sanitation: Roof Drainage

RUN-OFF FROM VERTICAL WALLS

It is generally appreciated that the run-off from vertical walls bounding a roof can add to the flow loading of gutters and rainwater pipes and therefore, some allowance should be made. Many factors are involved, but, in the absence of more precise data, one half of the area of vertical wall contributing, should be added to the roof area to be drained. If the roof is bounded by two or more walls, take half the area of the wall presenting the greatest projected area in elevation.

CARRYING CAPACITIES

Most of the design data presented by various authorities is based on gutters being fixed level. A gutter fixed with a fall of only 1 in 600(1mm in 0.6m) will carry up to 40 per cent more than when fixed level and the increased flow will assist in preventing silting. It is recommended that flow capacities of level gutters be used for design and the increased flow rate be regarded as a safety margin. Although level gutters should empty themselves via the outlet, a slight fall is recommended, particularly for eaves gutters, because of the adverse effect of disturbance by ladders, settlement of the buildings, etc.

EXAMPLE

Calculate the size of eaves gutter and rainwater pipe to drain a roof 20m long × 2.5m wide (on plan) with roof pitch of 60 deg. and with gutters to fall approximately 1 in 600. Allow for 75mm of rain per house (assume fall allows for safety margin).

Effective roof area = Plan area

Plan area = $(20 \times 2.5) + (0.5 \times 1.44 \times 20)$

$= 50m^2 + 14.4m^2$

$= 64.4m^2$

Q = Rate of run-off in litres/sec.

$Q = \dfrac{AeI}{3600}$ where I is the design rate of rainfall in mm/hr

$= \dfrac{64.4 \times 75}{3600}$ (mm/hr)

$= 1.34$ litres/sec.

From Table 1 of BS.6367: 1983: flow capacities of standard eaves gutters, a 125mm true half-round gutter @1.37 litres/sec. Outlet size (Table 5 from BS.6367: 1983) for standard eaves gutters: Sharp edged — 75mm diameter, round edged — 63mm diameter — when outlet is positioned at one end.

GUTTER AND RAINWATER PIPE SIZING TABLES EXTRACTED FROM BS6367: 1983

Table 1

Nominal size gutter	True half-round gutter (1, 2 and 5)	Nominal half-round (segmental) gutter (3, 4 and 5)
mm	L/s	L/s
75	0.38	0.27
100	0.78	0.55
115	1.11	0.78
125	1.37	0.96
150	2.16	1.52

Pitch o	Cos	Pitch o	Cos	Pitch o	Cos
5	.9962	30	.8660	55	.5736
10	.9848	35	.8192	60	.5000
15	.9659	40	.7660	65	.4226
20	.9397	45	.7071	70	.3420
25	.9063	50	.6428	75	.2588

To calculate true area of roof given plan and pitch

= plan area $\times \dfrac{1}{Cos\ A}$

Pitch o	Tan
55	1.428
60	1.732
65	2.145
70	2.747
75	3.732

Required for calculations for roof pitches over 50° (See detail 61)

Table 2

Half round gutter size	Sharp (SC) or round cornered (RC) outlet	Outlet at end of gutter	Outlet not at end of gutter
mm	—	mm	mm
75	SC	50	50
	RC	50	50
100	SC	63	63
	RC	50	50
115	SC	63	75
	RC	50	63
125	SC	75	89
	RC	63	75
150	SC	89	100
	RC	75	100

Rainwater pipe sizes (minimum nominal diameter)

Flow capacities of level gutters, outlet at end
1 Asbestos cement to BS569: 1973
2 Pressed steel to BS1091: 1963 (PS)
3 Aluminium to BS2997: 1958 (Al)
4 Cast iron to BS460: 1964 (CI)
5 uPVC to BS4576: Part 1:1970

ADJUSTMENTS

Effects of angles. If a length of eaves gutter includes an angle, the flow in the gutter will be impeded, and its capacity reduced. From Table 3 of BS6367, the capacity of standard eaves gutter should be reduced by the appropriate amount depending on the proximity of the outlet in relation to the angle.

Type of angle	Reduction factor	
	Angle less than 2m from outlet	Angle between 2m & 4m from outlet
Sharp corner	0.80	0.90
Round Corner	0.90	0.95

+ 40 per cent (f = 0.71)

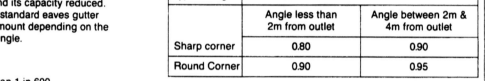

Falls not less than 1 in 600

Sloping gutters up to 6m long   **−10 per cent (f = 1.1)**

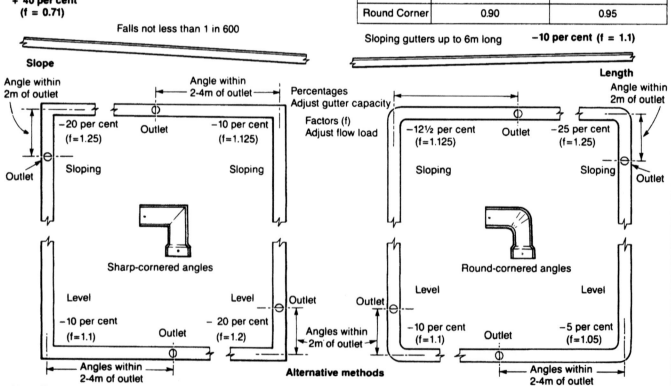

Note: These adjustments can be made to the flow load rather than to the gutter capacity, if preferred. This will assist in finding the gutter size and spacing of the outlets by reference to Table 1, for example, instead of increasing the gutter capacity say by 40%, multiply the flow load by 100 or 0.71 (f).

Sanitation: Roof Drainage

OUTLET FLOW PATTERNS

GUTTER CAPACITIES

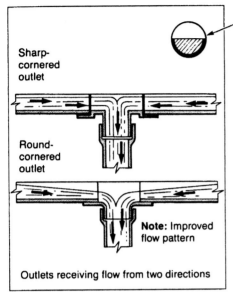

Sharp-cornered outlet

Round-cornered outlet

Note: Improved flow pattern

Outlets receiving flow from two directions

BS6367: 1983 states that both rectangular and triangular gutters may be considered as special types of trapezoidal gutter; and can therefore be designed in the same way as trapezoidal gutters. The BS acknowledges that the fall to which a gutter is laid will be dependent upon certain considerations — the material of which the gutter is formed, and structural reasons. It suggests that gutters can be laid flat, particularly in the form of prefabricated gutters of pressed steel and aluminium. It is beneficial, however, if sufficient fall is provided to prevent ponding. Where gutters are laid to fall, it has been found that the flow capacity of the gutters is increased. This, however, is seen as an additional margin of safety.

Flow capacities of level trapezoidal gutters complying with BS569 and discharging freely.

Q = litres per second
A = cross-sectional area in mm²
R.W. = rainwater

Examples
1. Calculate capacity of true half-round 100mm gutter using $Q = 0.0016\ A^{1.25}$
Cross-sectional area of gutter $= \dfrac{\pi d^2}{4} \div 2$

Therefore A = 3925mm²
$Q\ (l/s) = 0.0016 \times 3925^{1.25} \div 60 = 0.83\ l/s$
(Compared answer with Table 1)
2. Calculate size of gutter and rainwater pipe for roof 10m (ridge to eaves) × 50m, pitch 30° and gutters not to fall less than 1 in 600. The pitch if less than 50° therefore flow load per metre of eaves = 1.25 × 10(m) ÷ 60(s) = 0.2 l/s (see Sheet 65). Adjust for slope of 1 in 600 = 0.2 × $\dfrac{100}{140}$ (or × 0.71) = 0.14 L/s

By reference to Table 1, it can be seen that a 125mm true HR gutter could cope with 10m of eaves and therefore would require outlets spaced maximum 20m apart. Obtain rainwater pipe size from Table 2.

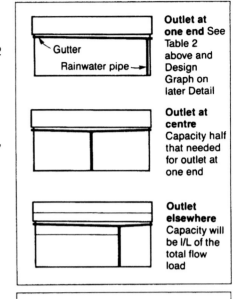

Outlet at one end See Table 2 above and Design Graph on later Detail

Outlet at centre Capacity half that needed for outlet at one end

Outlet elsewhere Capacity will be l/L of the total flow load

Trapezoidal
75mm Freeboard

Type of gutter (dimensions in mm)	Flow capacity (litres/sec)	Depth of Flow Yc at downstream end of gutter (mm)
Valley Gutters		
406×127×254	8.7	47
457×127×152	6.8	51
610×152×229	12.7	60
Northlight valley gutter		
457×152×102	7.6	64
Boundary wall gutter Pattern 1		
279×127×178	6.0	47
305×152×229	9.7	55
457×152×305	13.4	56
Pattern 2		
559×152×406	17.3	55

Note
Flow capacities include allowance for freeboard

S.I. METRIC UNITS

Taking theory of rectangular weir flow – flow rate = mean velocity of flow × effective cross-sectional area of weir i.e. $Q = V \times A$ from which

$Q = Cd \times 2/3 \sqrt{2gh} \times (B \times h)$

Allowing for circular weir and values to mm units

$Q = 0.64 \times 2/3 \sqrt{2} \times 9.8 \times \dfrac{h}{1000} \times$

$\dfrac{\pi D}{1000} \times \dfrac{h}{1000}$

$= 0.43 \sqrt{\dfrac{19.6}{1000}} \times \sqrt{h^3} \times \dfrac{3.14D}{10^6}$

$= 0.43 \sqrt{19.6} \times \sqrt{\dfrac{1}{1000}} \times \sqrt{h^3} \times$

$3.14 \times \dfrac{D}{10^6}$

$= \dfrac{0.43 \times 4.4 \times 3.14}{31.6} \times \sqrt{h^3} \times \dfrac{D}{10^6}$

$Q = \dfrac{D \sqrt{h_o^3}}{5 \times 10^6}$ (cubic metres per second)

(i)   (5.3 rounded off to 5)

Allowing for 75 mm of rain per hr. and seconds to hours

$Q \times \dfrac{75}{1000} = \dfrac{D_o \times h_o^3 \times 60 \times 60}{5 \times 1000 \times 1000}$

Let $Q \times \dfrac{75}{1000} =$

roof area in square metres (RA)

Therefore RA =

$\dfrac{1000 \times D_o \times \sqrt{h_o^3} \times 60 \times 60}{75 \times 5 \times 1000 \times 1000}$

$= 9 D_o \sqrt{h_o^3} - 1000$

(ii)   (9.6 taken as 9)

Flat roof application (no gutters)

Limiting head (h_o) to $\dfrac{D_o}{4}$ to avoid vortex

Therefore

$RA = 9D_o \sqrt{\left(\dfrac{D_o}{4}\right)^3} - 1000$

$= \dfrac{9 \sqrt{D_o^2 \times \dfrac{D_o}{4} \dfrac{D_o}{4} \dfrac{D_o}{4}}}{1000} = \dfrac{9\sqrt{\dfrac{D_o^5}{64}}}{1000}$

$= \dfrac{\dfrac{9}{1} \times \dfrac{1}{8} \times \sqrt{D_o^5}}{1000}$

Therefore $RA = \dfrac{\sqrt{D_o^5}}{1000}$

(assumes that $\dfrac{9}{1} \times \dfrac{1}{8}$ cancel)

or $D_o = \sqrt[5]{\left(\dfrac{RA}{0.001}\right)^2}$

(iii)

Outlets from flat roofs (no gutters)

Outlets from large gutters or small cesspools, the head over outlet can be allowed to increase to

$\dfrac{D}{3}$ (to avoid a vortex)

Taking expression

(ii) $RA = 9D_o \sqrt{h_o^3} - 1000$

and replacing h_o with $\dfrac{D_o}{3}$

Therefore $RA = \dfrac{9 D_o \sqrt{\dfrac{D_o^3}{3}}}{1000}$

$= \dfrac{9 \sqrt{D_o^2 \times \dfrac{D_o}{3} \times \dfrac{D_o}{3} \times \dfrac{D_o}{3}}}{1000} = \dfrac{9\sqrt{\dfrac{D_o^5}{27}}}{1000}$

$= \dfrac{\dfrac{9}{1} \times \dfrac{1}{5.2} \times \sqrt{D_o^5}}{1000}$

Therefore $RA = \dfrac{1.7 \sqrt{D_o^5}}{1000}$

or $D_o = \sqrt[5]{\left(\dfrac{RA}{0.0017}\right)^2}$

(iv)

For large gutter application

It can be shown by hydraulic principles that for any given roof area, if the outlet takes the form of a properly shaped transition piece, then some reduction of the diameter of the rainwater pipe compared with the diameter of the outlet is permissible. The following expression can be used –

$RA = 0.0032 \sqrt{D^5_{RWP}}$ or $D_{RWP} =$

$\sqrt[5]{\left(\dfrac{RA}{0.0032}\right)^2}$

(v)

Note
for tapered inlets only

N.B. Owing to the problem of assessing the correct amount of rain to allow for, exact mathematical accuracy is not an important factor

---

NOTES

Outlets should be designed to receive discharges at such a rate as to enable the outlet to discharge as a circular weir.

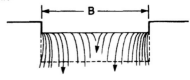

Rectangular weir

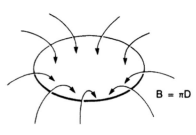

$B = \pi D$

Circular weir

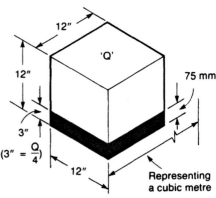

$(3'' = \dfrac{Q}{4})$

Representing a cubic metre

Flat roof application (no gutters)

Rainwater pipes with tapered inlets

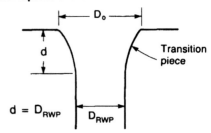

$d = D_{RWP}$ D_{RWP}

Transition piece

D_{RWP} = Diameter of rainwater pipe

METRIC UNITS AND LETTERS AND SYMBOLS USED

Q	Discharge rate in m^3/sec.	g	Gravitational acceleration 9.8 m/s^2	D	Outlet diameter in metres
V	Velocity in metres/sec.	h	Depth of water at outlet in m	D_o	Outlet diameter in millimetres 3.14
A	Cross-sectional area in m^2	h_o	Depth of water at outlet in mm	RA	Roof area in square metres
Cd	Co-efficient of discharge (0.64)	B	Mean breadth of weir in metres		

Sanitation Roof Drainage

IMPERIAL UNITS

Taking theory of rectangular weir flow – flow rate = mean velocity of flow × effective cross-sectional area of weir i.e. Q = V × A from which
$Q = Cd \times 2/3 \sqrt{2gh} \times (B \times h)$
Allowing for circular weir and values to inch units

$$Q = 0.64 \times 2/3 \sqrt{2 \times 32} \times$$

$$\frac{h}{12} \times \frac{\pi D}{12} \times \frac{h}{12}$$

$$= 0.43 \sqrt{\frac{64h}{12}} \times \frac{3.14\ Dh}{144}$$

$$= 0.43 \sqrt{5.3h^3} \times \frac{3.14 D}{144}$$

$$= \frac{0.43 \times 2.3 \times 3.14}{144} \times D \sqrt{h^3}$$

$$Q = 0.022\ D_o \sqrt{h_o^3}$$
(i)   (cubic feet per second)

Allowing for 3 in. of rain per hour and seconds to hours

$$\frac{Q}{4} = 0.022\ D_o \sqrt{h_o^3} \times 60 \times 60$$

Let $\frac{Q}{4}$ = roof area in square feet (RA)

Therefore RA =
$$4 \times 0.022 \times D_o \sqrt{h_o^3} \times 60 \times 60$$

$$= 320\ D_o \sqrt{h_o^3}$$
         (316.8 rounded to 320)
(ii)   **Flat roof application** (no gutters)

Limiting head (h_o) to $\frac{D_o}{4}$ to avoid a vortex

$$RA = 320\ D_o \sqrt{\frac{D_o}{4}}^{\ 3}$$

$$= 320 \sqrt{D_o^2 \times \frac{D_o}{4} \times \frac{D_o}{4} \times \frac{D_o}{4}}$$

$$= 320 \sqrt{\frac{D_o^5}{64}}$$

$$= 320 \times 1/8 \times \sqrt[5]{D_o^5}$$

Therefore $RA = 40 \sqrt{D_o^5}$

$$\text{or } D_o = \sqrt{\frac{RA^2}{40}}$$

(iii)

     Outlets from flat roofs (no gutters)

Outlets from large gutters or small 'cesspools', the head over the outlet can be allowed to increase to

$\frac{D}{3}$ (not $\frac{D}{4}$ as for flat roofs)

Taking expression
(ii) RA = $320\ D_o \sqrt{h_o^3}$
and replacing h_o with $\frac{D_o}{3}$

$$RA = 320\ D_o \sqrt{\frac{D_o}{3}}^{\ 3}$$

$$= 320 \sqrt{D_o^2 \times \frac{D_o}{3} \times \frac{D_o}{3} \times \frac{D_o}{3}}$$

$$= 320 \sqrt{\frac{D_o^5}{27}}$$

$$= 320 \times \frac{1}{5.2} \sqrt{D_o^5}$$

Therefore $RA = 60 \sqrt{D_o^5}$
         (61.5 rounded off to 60)

$$\text{or } D_o = \sqrt{\frac{RA^2}{60}}$$

(iv)

     For large gutter application

It can be shown by hydraulic principles that for any given roof area, if the outlet takes the form of a properly shaped transition piece, then some reduction of the diameter of the rainwater pipe compared with the diameter of the outlet is permissible. The following expression can be used —

$$RA = 120 \sqrt{D^5_{RWP}} \text{ or } D_{RWP} =$$

$$\sqrt[5]{\left(\frac{RA}{120}\right)^2}$$

(v)

Note
for tapered inlets only

N.B. Owing to the problem of assessing the correct amount of rain to allow for exact mathematical accuracy is not an important factor

IMPERIAL UNITS AND LETTERS AND SYMBOLS USED

Q	Discharge rate in cu. ft./sec.	g	Acceleration (gravity) 32 ft/s²	D	Outlet diameter in feet
V	Velocity in feet/second	h	Depth of water at outlet in ft.	D_o	Outlet diameter in inches 3.14
A	Cross-sectional area in sq. ft.	h_o	Depth of water at outlet in ins.	RA	Roof area in square feet
Cd	Co-efficient of discharge (0.64)	B	Mean breadth of weir in ft.		

Sizing of Rainwater Goods (4)

SIZING OF EAVES GUTTERS AND OUTLETS

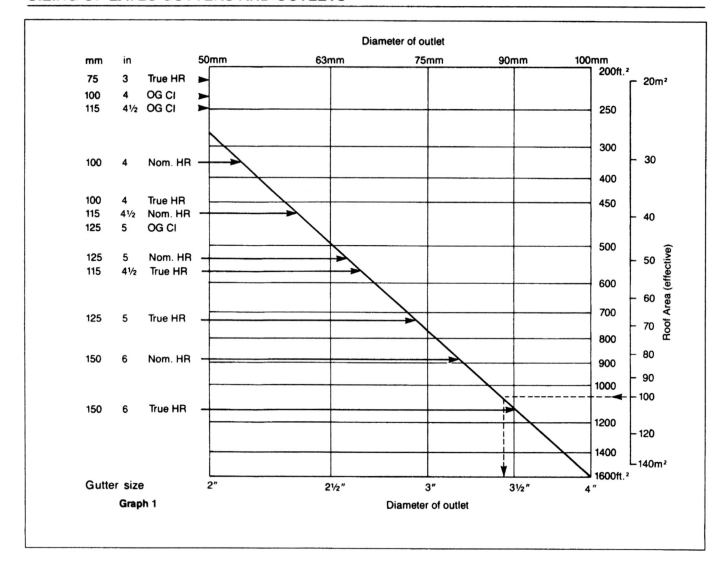

Graph 1

Graph 1, on which suitable gutter outlet sizes are also indicated, is based on flow capacities for level gutters with an outlet at one end.

Example for pitched roof of 100m² of effective roof area, with eaves gutter (actual — not plan). Size of outlet and rainwater pipe:- 90 mm (non-tapered nozzle). Size of eaves gutter:- the graph indicates that a 150mm true half-round gutter will be suitable.

HR = Half round
OG CI = Cast iron with ogee curve

Sanitation: Roof Drainage

OUTLET AND RAINWATER PIPE SIZING

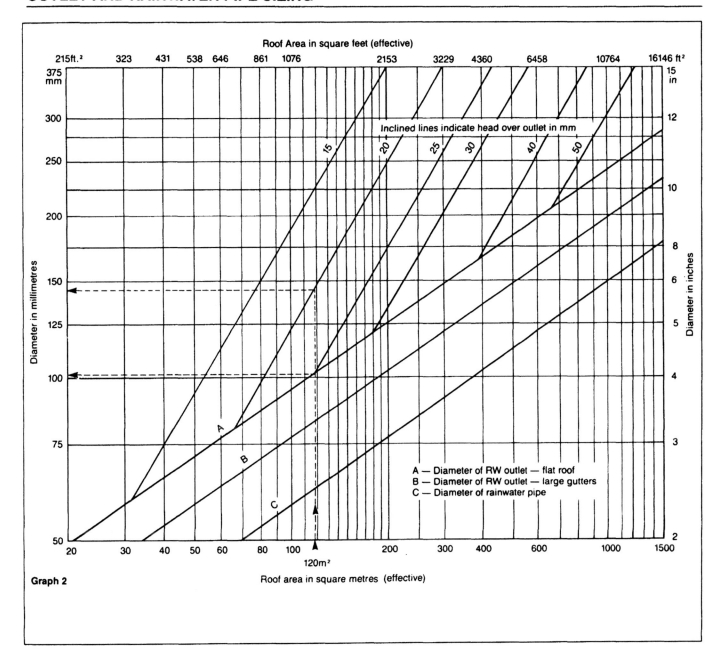

Graph 2

Roof area in square metres (effective)

Graph 2 has been drawn to include formulae (ii), (iii) and (iv) (see previous Detail Sheet). Problems requiring the application of these basic hydraulic formulae can thus be solved at a glance (the same applies to Graph 1), with the minimum of calculations.

Graph 2 has been plotted to determine suitable diameters for outlets for flat roofs not provided with gutters or cesspools (A), and also outlets for those roofs provided with gutters (B). In cases where it is required to limit the depth of the water at the outlet to a flat roof, the necessary data has been incorporated. If the gutter outlets are bell-mouthed (using tapered transition piece), then reference to line C will indicate the possible reduction in rainwater pipe size. Where the standard outlet nozzle is used, the rainwater pipe should be the same diameter as the outlet. Example for

120m² of flat roof (no gutters). Line A — outlet:- 100mm. Assuming tapered inlet — line C — RW pipe:- 63mm. If depth of water limited to 20mm:- 150mm outlet and 63mm RW pipe with tapered inlet.

The above graphs are reproduced with the kind permission of the Institute of Plumbing.

Sizing of Industrial Gutters

INTRODUCTION

For structural reasons little or no fall can usually be provided and it can be said that any built-in fall merely assists in draining the gutter. Therefore, with the numerical value of incline (I) being zero, the normal hydraulic formulae cannot be used to determine the dimensions of level gutters. If conditions existing at the outlet or point of free discharge are used, a rational approach is possible if a fundamental flow formula is applied. It can be shown that the depth of water at the point of free discharge, i.e. maximum flow for a given depth will be two thirds of the depth upstream but to provide a safety margin, design for depth of flow at 'still' end to be twice the depth at the point of free discharge. In addition, allow for 'freeboard'* of 50 to 60 mm and minimum gutter width of 250 mm.

*Freeboard — height above water level to top of gutter edge when flowing at maximum rated depth.

S.I. METRIC UNITS

Basic expression for flow in gutter at point of free discharge: $Q = V \times A$

Q = Mean velocity × cross-section of water at outlet. Taking rainfall rate of 75 mm per hour and completing all conversions and using $Q = V \times A$

$$Q \text{ or } \frac{75 \times RA}{1000 \times 3600} = \frac{V \times A}{1000^2}$$

Therefore
$$RA \text{ (m}^2) = \frac{V \times A}{1000^2} = \frac{3600 \times 1000}{75}$$

and $RA = V \times A \times 0.05$ or $\dfrac{V \times A}{20}$

A in mm². V in m/s.

If $RA = \dfrac{V \times A}{20}$ then $V = \dfrac{RA \times 20}{A}$ and if

$V = \dfrac{\sqrt{h_o}}{11}$ then $\dfrac{RA \times 20}{A} = \dfrac{\sqrt{h_o}}{11}$ and

$$RA = \frac{A\sqrt{h_o}}{11 \times 20} = \frac{A\sqrt{h_o}}{220}$$

As $B \times h_o = A$, then $RA = \dfrac{B\sqrt{h_o^3}}{220}$

Alternative: $RA = 0.0045\, B\sqrt{h_o^3}$ or $h_o = $

$$\sqrt[3]{\left(\frac{RA}{0.0045B}\right)^2}$$

(ii)

If depth of flow is acceptable when:

$h_o = \dfrac{D_o}{3}$ and with the cross-sectional area being $A = B \times h_o$,

then $A = B \times \dfrac{D_o}{3}$

Formulae for cross-sectional area etc. of roof gutters

If $B = 2D_o$ (see iii),

then $A = 2D_o \times \dfrac{D_o}{3}$ and $A = \dfrac{2D_o^2}{3}$

(iv)

Mean velocity of flow for any given depth is given by:- V (m/s) $= 2/3 \sqrt{2gH}(m)$

Converting head to mm units, then

$$V = 2/3\sqrt{\frac{2 \times 9.8 \times H}{1000}} = 2/3\sqrt{\frac{19.6}{1000}}$$

$\times \sqrt{h_o} = 2/3\sqrt{0.0196} \times \sqrt{h_o} = $

$2/3 \times 0.14 \times \sqrt{h_o} = 0.09\sqrt{h_o}$ or $\dfrac{\sqrt{h_o}}{11}$

(i)

For any given conditions of free flow at the outlet, the flow capacities of both gutter and outlet must be equal from which the following relationships will hold:

$$\frac{B\sqrt{h_o^3}}{220} = \frac{9D_o\sqrt{h_o^3}}{1000}$$

Therefore $D_o = \dfrac{1000B}{220 \times 9} = \dfrac{1000B}{1980}$

$\dfrac{1000}{1980}$ assumed to be ½

$D_o = \dfrac{B}{2}$ and $B = 2D_o$

(iii)

Note: Minimum width where access to the gutter is required: 250 mm

75 mm could be taken as 0.075m, instead of dividing by 1000.

3600 converts hr. to sec.

1000² converts m² to mm²

METRIC UNITS AND EXPLANATION OF LETTERS USED

Q	Discharge rate m³/time factor	H	Head in metres or mm	B	Mean breadth of gutter in mm
RA	Roof area in square metres	g	Acceleration (gravity) 9.8 m/s²	D_o	Outlet diameter in mm
V	Velocity in metres/time factor	h_o	Depth of water at outlet in mm		
A	Cross-sectional area in m² or mm²	m/s	Metres per second		

Sanitation: Roof Drainage

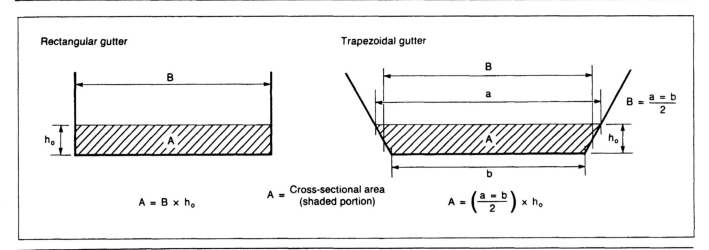

Rectangular gutter

Trapezoidal gutter

$B = \dfrac{a=b}{2}$

$A = B \times h_o$

$A = $ Cross-sectional area (shaded portion)

$A = \left(\dfrac{a=b}{2}\right) \times h_o$

IMPERIAL UNITS

Basic expression for flow in gutter at point of free discharge: $Q = V \times A$

Q = Mean velocity × cross-section of water at outlet. Taking rainfall rate of 3 in. per hour and completing all conversions and using $Q = V \times A$

Q or $\dfrac{1 \times RA}{4 \times 3600} = V \times \dfrac{A}{144}$

(converts sq. ft. to sq. in.)

Therefore RA (sq. ft.) =

$V \times \dfrac{A}{144} \times \dfrac{3600}{1} \times \dfrac{4}{1}$

and $RA = V \times A \times 100$ A in sq. in.
V in ft./sec.

Mean velocity of flow for any given depth is given by:- V (ft/s) = $2/3 \sqrt{2gH}$ (ft)

Converting head to inch units, then V =

$2/3\sqrt{\dfrac{2 \times 32 \times H}{12}} = 2/3\sqrt{\dfrac{64}{12}} \times \sqrt{h_o}$

$= 2/3 \sqrt{5.33} \times \sqrt{h_o} =$
$2/3 \times 2.3 \times \sqrt{h_o} = 1.6 \sqrt{h_o}$

1.53 rounded up to 1.6

(i)

If $RA = V \times A \times 100$ then

$V = \dfrac{RA}{A \times 100}$ and if $V = 1.6 \sqrt{h_o}$ then

$\dfrac{RA}{A \times 100} = 1.6 \sqrt{h_o}$ and

$RA = 160\, A \sqrt{h_o}$
As $B \times h_o = A$, then $RA = 160\, B \sqrt{h_o^3}$

(ii)

For any given conditions of free flow at the outlet, the flow capacities of both gutter and outlet must be equal from which the following relationships will hold:

$320\, D_o \sqrt{h_o^3} = 160\, B \sqrt{h_o^3}$

Therefore $D_o = \dfrac{160B}{320}$

$D_o = \dfrac{B}{2}$ and $B = 2D_o$

(iii)

If depth of flow is acceptable when:

$h_o = \dfrac{D_o}{3}$ and with the cross-sectional area being $A = B \times h_o$,

then $A = B \times \dfrac{D_o}{3}$

Formulae for cross-sectional area etc. of roof gutters

If $B = 2D_o$, (see iii), then

$A = 2D_o \times \dfrac{D_o}{3}$ and $A = \dfrac{2D_o^2}{3}$

(iv)

Note: Minimum width where access to the gutter is required: 10 in.

3″ is taken as ¼ ft.

3600 converts hr. to sec.

IMPERIAL UNITS AND EXPLANATION OF LETTERS USED

Q	Discharge rate cu. ft./time factor	V	Velocity in feet/time factor	ft/s	Feet per second
RA	Roof area in square feet	H	Head in feet or inches	B	Mean breadth of gutter in inches
A	Cross-sectional area in sq. ft. or sq. in.	g	Acceleration (gravity) 32 ft./s²	D_o	Outlet diameter in inches
		h_o	Depth of water at outlet in inches		

Worked Example Simple Domestic Arrangement

ROOF DATA

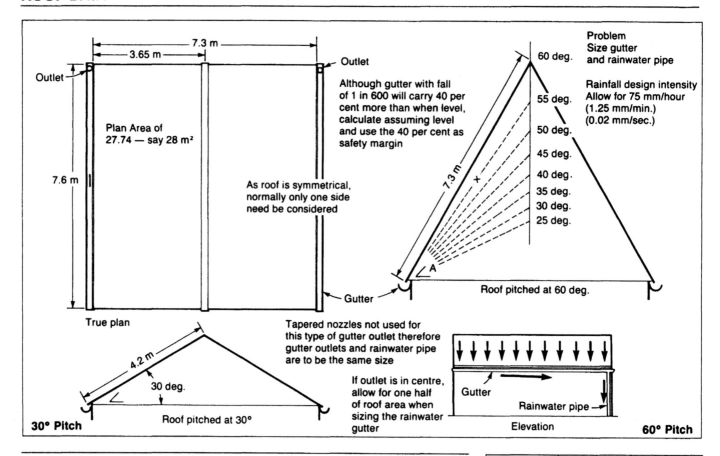

True plan

Plan Area of 27.74 — say 28 m²

7.6 m

3.65 m

7.3 m

Outlet

Outlet

Although gutter with fall of 1 in 600 will carry 40 per cent more than when level, calculate assuming level and use the 40 per cent as safety margin

As roof is symmetrical, normally only one side need be considered

Gutter

30° Pitch

4.2 m

30 deg.

Roof pitched at 30°

Tapered nozzles not used for this type of gutter outlet therefore gutter outlets and rainwater pipe are to be the same size

If outlet is in centre, allow for one half of roof area when sizing the rainwater gutter

Problem
Size gutter and rainwater pipe

Rainfall design intensity
Allow for 75 mm/hour (1.25 mm/min.) (0.02 mm/sec.)

60 deg.
55 deg.
50 deg.
45 deg.
40 deg.
35 deg.
30 deg.
25 deg.

7.3 m

x

A

Roof pitched at 60 deg.

Gutter

Rainwater pipe

Elevation

60° Pitch

SOLUTIONS

Using a selection of methods

Using Approved Document H3 method from the Building Regulations 1985

From Table 1 of the Approved Document Design area given as plan area of portion × 1.15
Therefore:
28 m² × 1.15 = 32.2 m² effective area

From Table 2 of the Approved Document A 100 mm diameter gutter, laid level up to 8 m in length: half-round in section with a sharp edged outlet only at one end, will serve an effective area of 37 m² with a minimum outlet of 63 mm diameter. (Flow capacity = 0.78 litres/sec.).

Using Approved Document H3 method from the Building Regulations 1985

From Table 1 of the Approved Document Design area = plan area × 2
Therefore:
28 m² × 2 = 56 m²

From Table 2 of the Approved Document A 125 mm diameter gutter will serve an effective area of 65 m² with a 75 mm diameter outlet at one end. (Flow capacity = 1.37 litres/sec.).

Note: Round edged outlets allow smaller down pipe size. The Approved Document also suggests: where there is a fall or the gutter has a section which gives it a larger capacity than a half-round gutter, it may be possible to reduce the size of the gutter and pipe. Paragraph 1.8 refers to relevant recommendations of BS.6367: 1983.

Angle of pitch	Length of pitched roof	Area: x × 7.6 m
deg.	x in m	m²
25	4.027	30.6
30	4.215	32.0
35	4.455	33.9
40	4.764	36.2
45	5.162	39.2
50	5.678	43.2
55	6.363	48.4

Sanitation: Roof Drainage

Rule of thumb for sizing rainwater pipes (useful for estimating)

Allow 650 mm² cross-sectional area of rainwater pipe per 9 m² of roof area*, therefore

$$\frac{\text{actual area (32 m}^2)}{9} = 3.6, \text{ then } \times 650$$

$$= 2340 \text{ mm}^2.$$

As the cross-sectional areas of 50 mm and 63.0 mm rainwater pipes are 2000 mm² and 3100 mm² (approximately) respectively, the designer would probably select the 63.0 mm size which is 'oversizing' but allows for a large safety factor.

Allow 650 mm² cross-sectional area of rainwater pipe per 9 m² of roof area*, therefore

$$\frac{\text{actual area (56 m}^2)}{9} = 6.2, \text{ then } \times 650$$

$$= 4030 \text{ mm}^2.$$

As the cross-sectional areas of 63.0 mm and 75 mm rainwater pipes are 3100 mm² and 4400 mm² (approximately) respectively, the designer would probably select the 75 mm size which is 'oversizing' but allows for a large safety factor.

A simple calculation method for approximate sizing of rainwater pipes can be used if necessary, but designers would no doubt prefer to use tables or graphs for speed and simplicity.

Diameter of rainwater pipe in mm =

$$\sqrt[5]{\left(\frac{RA}{0.0016}\right)^2}$$

RA = actual roof area in m²

Therefore $D = \sqrt[5]{\left(\frac{32}{0.0016}\right)^2}$

$$= \star \; 400,000,000 = 53 \text{ mm}$$

Once again it would appear that to select 50 mm rainwater pipe is slightly 'undersizing', but 63.0 mm allows safety margin.

Diameter of rainwater pipe in mm =

$$\sqrt[5]{\left(\frac{RA}{0.0016}\right)^2}$$

RA = Actual roof area in m²

Therefore $D = \sqrt[5]{\left(\frac{56}{0.0016}\right)^2}$

$$= \sqrt[5]{1,225,000,000} = 66 \text{ mm}$$

Again it would appear that to select 63.0 mm rainwater pipe is slightly 'undersizing', but 75 mm allows safety margin.

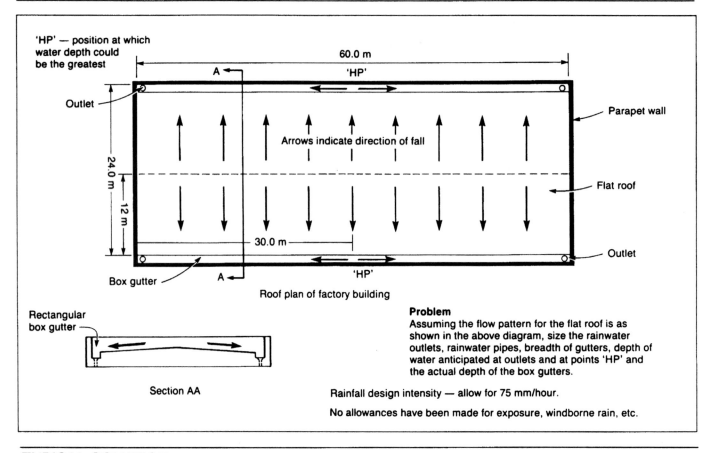

'HP' — position at which water depth could be the greatest

Outlet

60.0 m

'HP'

A

24.0 m

12 m

Arrows indicate direction of fall

Parapet wall

Flat roof

30.0 m

Outlet

Box gutter

A

'HP'

Roof plan of factory building

Rectangular box gutter

Section AA

Problem
Assuming the flow pattern for the flat roof is as shown in the above diagram, size the rainwater outlets, rainwater pipes, breadth of gutters, depth of water anticipated at outlets and at points 'HP' and the actual depth of the box gutters.

Rainfall design intensity — allow for 75 mm/hour.

No allowances have been made for exposure, windborne rain, etc.

TYPICAL SOLUTION

(i) Sizing outlets
Actual plan of roof area:- 60 m × 24 m = 1440 m². Allowing for four outlets and equal distribution of run-off (in theory), effective area of roof per outlet (RA) = 1440 ÷ 4 = 360 m². Using expression based on the theory of rectangular weir flow converted to circular weir flow and limiting depth of flow to avoid vortex then —
Diameter of outlet (D_o) =

$$\sqrt[5]{\left(\frac{RA}{0.0017}\right)^2}$$

Therefore D_o $\sqrt[5]{\left(\frac{360}{0.0017}\right)^2}$

= 135 mm. Select 150 mm as the most appropriate commercial size and rainwater pipe must be the same size unless a bellmouth entry is used. (Fig. 1). For rule of thumb (box gutter outlets) use 350 mm of cross-sectional area of outlet per 9 m² of roof area = 14,000 mm² — this method also indicates an outlet size between 125 and 150 mm (see Table 1).

(ii) Sizing rainwater pipes
Allowing for introduction of bellmouth entry (Figure 1) to rainwater pipe, then

$$D_{RWP} = \sqrt[5]{\left(\frac{RA}{0.0032}\right)^2}$$

therefore D_{RWP} $\sqrt[5]{\left(\frac{360}{0.0032}\right)^2}$

= 105 mm.

Theoretically 100 mm is slightly undersized and could be accepted on the grounds of economy, but some engineers may select 125 mm, which would provide a large amount of reserve capacity which may be necessary for windborne rain.

(iii) Calculating breadth of gutters

Assuming a depth of flow of $\frac{D_o}{3}$ is acceptable to avoid a vortex and the flow capacity of outlet must equal flow capacity of gutter, then

$\frac{B\sqrt{h_o^3}}{220} = \frac{9 D_o \sqrt{h_o^3}}{1000}$ which simplifies

to $D_o = \frac{1000B}{1980}$ and $D_o = \frac{B}{2}$ or B = 2 D_o, therefore breadth of gutter is B = 2 × 135 = 270 mm but 300 mm is a more practical dimension.

Sanitation: Roof Drainage

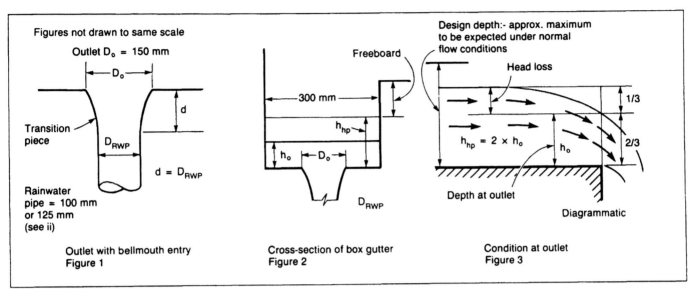

Figures not drawn to same scale

Outlet D_o = 150 mm

Transition piece

Rainwater pipe = 100 mm or 125 mm (see ii)

$d = D_{RWP}$

Outlet with bellmouth entry
Figure 1

Freeboard

300 mm

h_{hp}

h_o — D_o

D_{RWP}

Cross-section of box gutter
Figure 2

Design depth:- approx. maximum to be expected under normal flow conditions

Head loss

1/3

$h_{hp} = 2 \times h_o$

h_o

2/3

Depth at outlet

Diagrammatic

Condition at outlet
Figure 3

(iv) Calculating approximate depth of flow at outlets

If depth of flow (h_o) should not exceed

$\frac{D_o}{3}$ and $A = B \times h_o$, then

cross-sectional area (A) $= B \times \frac{D_o}{3}$.

If $B = 2D_o$ then $A = 2D_o \times \frac{D_o}{3}$

or $\frac{2 D_o^2}{3}$ therefore $A = \frac{2 \times 150^2}{3}$

= 15,000 mm². Allowing a breadth of 300 mm, depth at outlet end is 50 mm which is acceptable. (Maximum being

$\frac{D_o}{3} = \frac{150}{3} = 50$ mm). To check depth:-

RA $= 0.0045 B\sqrt{h_o^3}$ which gives value of h_o as approximately 42 mm (Figure 2).

(v) Approximate depth of flow at maximum run-off (about points 'HP'). See Figures 2 and 3

Assuming breadth of gutter remains constant throughout its length, depth of flow (in theory) at points 'HP' should be taken as twice the depth of flow at the outlet end therefore depth is between say, 85 mm and 100 mm.

(vi) Actual depth of box gutter

(Depth at outlet × 2) + freeboard allowance of 50 to 60 mm therefore actual depth could be 150 mm (Figures 2 and 3).

Table for easy reference to cross-section areas

Rainwater outlets and pipes			
Diameter		Cross-sec.area Approx	
in	mm	in²	min²
2	50	3.14	2,000
2½	63	4.9	3,100
3	75	7.0	4,400
3½	89	9.6	6,200
4	100	12.6	7,900
5	125	19.6	12,300
6	150	28.3	17,700

Table 1

RWP = rainwater pipe
D_o = diameter of outlet
D_{RWP} = diameter of rainwater pipe
B = breadth of gutter
h_o = depth of water at outlet
h_{hp} = depth of water at HP
A = cross-sectional area (all in mm)

Snowfall. The run-off resulting from falling or accumulated snow is less than the lowest design rate therefore no special considerations are necessary other than:-
(1) Snowboards to bridge gutters and prevent them being blocked by snow especially if frozen.
(2) Use of snowguards at eaves where sliding snow may cause damage to gutters, conservatory, etc.

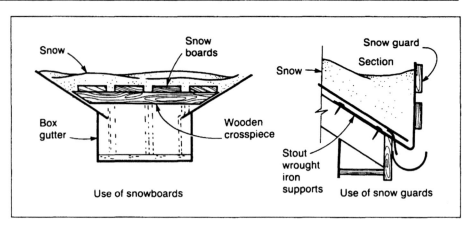

Use of snowboards

Use of snow guards

Provision of weir overflow. Where overflowing along the length of the gutter cannot be tolerated, it may be necessary to provide a weir overflow at the end of the gutter so that excess flow can be discharged clear of the building. Box receivers can also be designed as overflows.

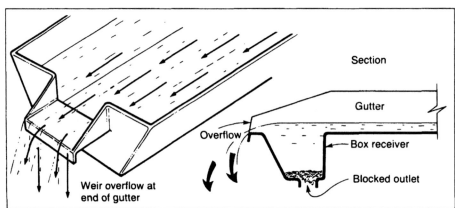

Weir overflow at end of gutter

Outlets provided with gratings. Outlets may have to be provided with gratings which could cause considerable restriction to the flow of water. One solution is to calculate the discharge capacity as normal and multiply by the ratio:-

$$\frac{\text{Area of perforations in the grating}}{\text{Total area of grating}}$$

Another approach is to find the equivalent horizontal roof area which can be drained to flat-grated outlet by using the expression:

$$RA(m^2) = \frac{Cd\, a\, \sqrt{h}}{3r}$$

where RA = roof area; Cd = 0.6 for bars with sharp edges or 0.8 for bars with rounded edges; a = total area of slots in mm^2; h = head over grating in mm and r = rainfall allowance in mm/hour. For domed gratings — refer to manufacturers' literature.

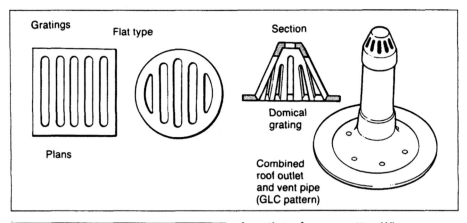

Combined roof outlet and vent pipe (GLC pattern)

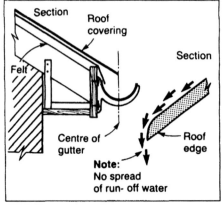

Note: No spread of run-off water

Location of eaves gutter. Where practicable, the gutter should be fixed centrally under the roof edge and close to the roof edge. The roofing felt should extend to just below the top edge of the gutter to prevent wind blowing rain behind the gutter. Gutter should not be permitted to tilt. Best roof edge is rounded at top and with sharp lower corner.

Sanitation: Roof Drainage

Box receivers. The box type receiver should be at least as wide as the maximum gutter width and should be long enough to prevent the flow from overshooting the box. Such receivers are best fixed at such a height that the overflow weir is level with the sole of the gutter. See CP308 for data for calculating dimensions.

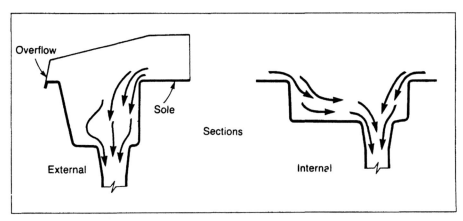

Rainwater inlets to discharge pipes. Roof and rainwater outlets may be connected directly to a discharge pipe(s) provided that (i) the drainage system is combined, not surcharged and would not be surcharged by the additional flow; (ii) roof area is not more than 186 m²; the stack is minimum 100 mm and allowance has been made for the extra flow; (iii) normal ventilation of the stack is maintained in case the outlet becomes blocked — this can be either by trapped or untrapped branch or G.L.C. pattern as above; (iv) outlet for rainwater (unless trapped) and outlet of vent terminates in safe position.

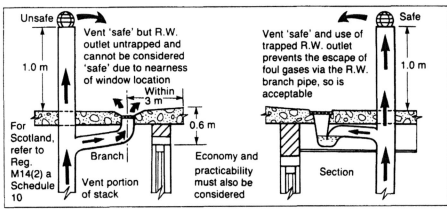

Positions where RWP's can be omitted. It is suggested in BS Code of Practice 6367:1983 that rainwater pipes may be omitted from roofs at any height with an area of less than 6 m² and provided that no other surface drains onto it. Consideration may also be given to the omission of rwp's from very tall structures with larger roof area.

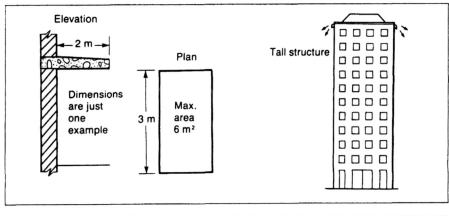

Access. One or more access points should be provided at appropriate points on horizontal runs and on long vertical runs. Access at the foot of each stack and at bends is particularly important.

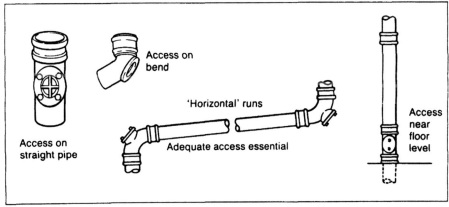

Balcony drainage unit

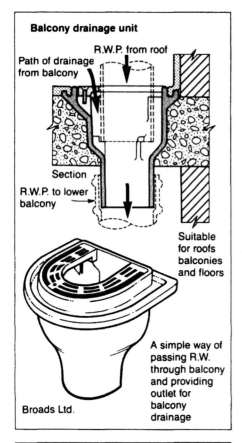

Path of drainage from balcony

R.W.P. from roof

Section

R.W.P. to lower balcony

Suitable for roofs balconies and floors

A simple way of passing R.W. through balcony and providing outlet for balcony drainage

Broads Ltd.

Lower termination of rainwater pipes. Where there is no alternative to a r.w.p. discharging on to a lower roof (or onto a paved area), a rainwater shoe should be fitted to divert the water away from the building. In certain cases it may be necessary to reduce splashing by fitting special shoes. It is sometimes advisable to reinforce the flat roof covering around the area of impact because of the possibility of localised wear. Where a r.w.p. discharges into a gully, it should terminate below the grating, preferably by the use of a back inlet gully. For internal rainwater pipes, the discharge can be via a gully or direct to a drain. Building Regulations require means of access as are necessary to permit internal cleansing.

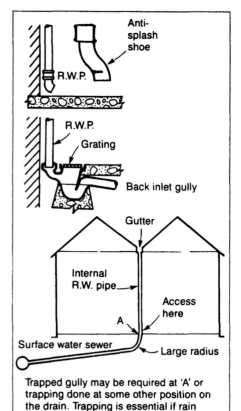

Anti-splash shoe

R.W.P.

R.W.P.

Grating

Back inlet gully

Gutter

Internal R.W. pipe

Access here

A

Surface water sewer

Large radius

Trapped gully may be required at 'A' or trapping done at some other position on the drain. Trapping is essential if rain water discharges into a foul sewer.

Roof outlets

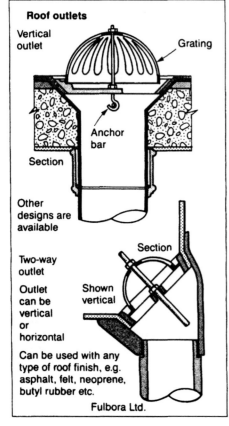

Vertical outlet

Grating

Section

Anchor bar

Other designs are available

Two-way outlet

Outlet can be vertical or horizontal

Shown vertical

Section

Can be used with any type of roof finish, e.g. asphalt, felt, neoprene, butyl rubber etc.

Fulbora Ltd.

Warning of blockage. Internal rainwater pipes should be able to withstand the head of water likely to occur due to a blockage, but in order to indicate that the pipe is in danger of being filled up to roof level, warning pipes should be provided at a height of 6m maximum above the points where blockage is likely, e.g. foot of stock, bends. Warning pipes to be minimum size 19mm and to take an upward direction from the stack and discharge in a visible position.

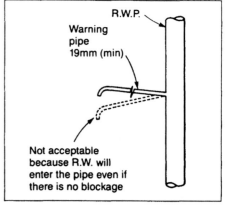

R.W.P.

Warning pipe 19mm (min)

Not acceptable because R.W. will enter the pipe even if there is no blockage

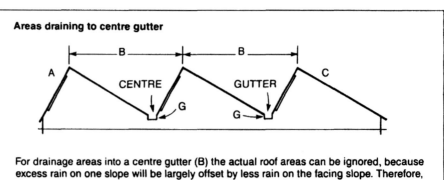

Areas draining to centre gutter

A B B C

CENTRE GUTTER

G G

For drainage areas into a centre gutter (B) the actual roof areas can be ignored, because excess rain on one slope will be largely offset by less rain on the facing slope. Therefore, the two slopes draining to (G) can be taken as Plan area for the purposes of design. For areas A and C, there is no compensating factor and therefore actual roof area is used.

Sanitation: Roof Drainage

Minimum number of gutter outlets and layout of flat roofs. Where overflowing along the length of a gutter cannot be tolerated, a minimum of two r.w.ps is desirable or the system should be designed for the highest rate of rainfall. It also may be necessary to provide a weir overflow at the end of the gutter. Some 'free-board' over the depth required to give the design capacity, is desirable as a safeguard against overflowing due to wave action caused by wind. Flat roofs may be designed to drain in several ways, e.g. (a) towards the 'outlet' edge with no gutters or outlets formed in the roof construction, (b) towards channels or gutters within the roof area, (c) to outlets (no gutters) within the roof area. An economic scheme should include few outlets, but this cannot always be attained.

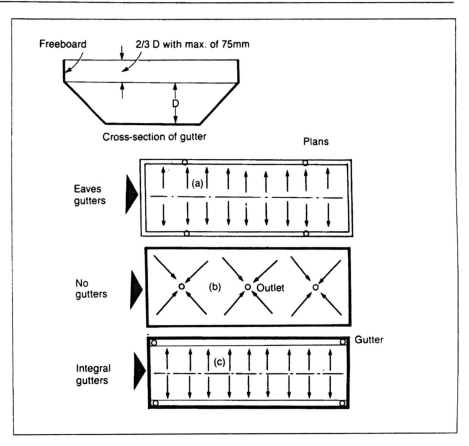

Encased pipes. Rainwater pipes may be encased in concrete columns or masonry walls provided that the requirements of the Building Regulations are met and precautions are taken to protect certain materials against chemical reaction with cement. It is important to provide appropriate access points. Approved Document H3 of the Building Regulations 1985 refer to size, materials, strength, durability, jointing, bends, thermal movement, access to pipe and internal cleansing.

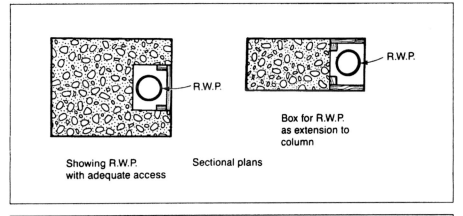

Subsidence. Undermined ground is displaced downwards and inwards towards the centre of the subsided area. The effect upon roof gutters may be to increase or decrease their gradients by perhaps 1 in 50, in extreme cases. The amounts of strain, etc., likely to occur at various points cannot be predicted with reasonable accuracy and advice should be sought from N.C.B. surveyor or other mineral surveyor.

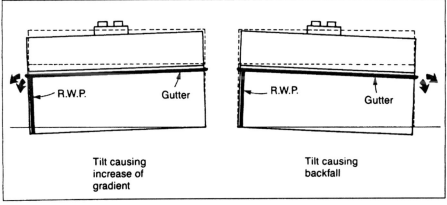

Flat roofs

Flat roofs should be designed to avoid ponding, except that some degree of temporary accumulation of water during heavy storms may be permitted where the roof covering is specially designed to remain watertight under such conditions. Each case needs individual consideration.

Horizontal rainwater pipes

These should be designed using normal hydraulics formulae for uniform flow. Suitable formulae are Crimp & Bruges' or Manning's using a roughness factor (n) of 0.012, or the Colebrook-White formula using a factor of 0.6 e.g. Crimp & Bruges:- $V = 83 \sqrt[3]{r^2} \sqrt{S}$ where V= velocity in m/s; r= hydraulic mean depth in metres; s = head/length or inclination. Manning:- $V = \dfrac{1}{n} \sqrt[3]{r^2} \sqrt{S}$.

Also Q= V × A therefore
$V = \dfrac{Q}{A}$ and $Q = \dfrac{1}{A} \dfrac{r^{2/3} S^{1/2}}{0.012}$.
Q= Flow in cumecs and
A= cross sec. area in m.

Inlets to rainwater pipes

The rainwater pipe inlet (or gutter outlet) should normally be so designed as to accept the flow without increasing the depth of flow in the gutter over that for free discharge. Where the rainwater pipe inlet is flush with the surface of a flat roof, the diameter of the inlet may be determined by a limitation on the maximum desirable depth of water allowable on the roof.

Location of rainwater pipes

Rainwater pipes may be fitted externally or internally, depending on design considerations and the occupational use of the building. It is possible to locate rainwater pipes in the most efficient locations by the full use of BS Code of Practice 6367: 1983 bearing in mind structural considerations and undergound drainage. With flat roofs, the structure may be affected by location of channels.

Parapet height and plan shape of roof

If the roof design includes parapets and sufficient upstand of asphalt or other roof covering is provided to prevent water affecting walls or from entering the building by way of tank rooms, roof lights or roof access staircases, there is, theoretically, no limit to the area that can be drained to one outlet. However, large areas draining to one outlet are undesirable because of the depth of water which can build up during a storm which may cause seepage through the roof due to flaws and/or bad workmanship.

Periodic inspection and cleaning

Gutters, rainwater pipes, outlets and particularly gratings, should be inspected and thoroughly cleaned once a year or more often if the building is in or adjoins an industrial area; also if it is near trees or may be subjected to extremes of temperature. All defects should be made good. Inspect gullies regularly and clean as often as necessary.

Protection of roof drainage outlets

Gratings are always a potential source of blockage and should be used only on outlets of 150mm nominal bore and above. The capacity of an outlet will generally be reduced if a protective grating is included. During the course of construction, outlets should be protected in order to prevent building materials and loose chippings from entering the rainwater pipe.

Run-off

Run-off should be calculated on the assumption that the roof surface is impervious. The rate of run-off is then the product of the effective roof area and the design rate of rainfall.

Testing

All work to be concealed should be tested before it is finally enclosed. Internal pipework should be tested for soundness, as should long flat runs of external pipework. Sealed rainwater pipes should have a pressure applied equal to 38mm water gauge* and after a period for stabilisation, should remain constant for 3 minutes (minimum) indicated on a manometer connected to a test plug via a branch of 'T' piece connector. The test should be applied using air or smoke from a smoke testing machine and introduced through the other branch of the 'T' piece. All nominally level gutters over walls and internal areas should be tested for leakage, after the gutter outlets have been blocked, by filling the gutter with water to the overflow level, if any, but otherwise to the lower level of the freeboard. It is important that all work should be continually inspected during installation. On completion, all rainwater pipes should be rodded through to ensure a clear bore.

*A head of water of 1 metre is approximately equal to 10 kN/m².

Sanitation: Roof Drainage

Vertical rainwater pipes
The rainwater pipes should be of the same nominal bore as the gutter outlet to which it is attached when designing as per Table 4 (Recommended minimum outlet sizes for eaves gutters) in BS Code of Practice 6367: 1983. As this procedure will result in pipes tending to flow full, the joint(s) between gutter outlet and rainwater pipe should be sealed. Standard offsets and shoes do not significantly restrict flow.

Walls draining to paved areas
In considering the drainage of paved areas, an allowance for run-off from vertical walls rising from the paved area will rarely be necessary, but, if the paved area is so depressed or situated that flooding might occur that cannot be tolerated, then the allowance for run-off from vertical surfaces should be added to the run off for the paved area.

Walls of court yards
For an enclosed area, only the run-off from the plan area of the court need be considered unless the surrounding walls are of unequal heights, in which case an additional run-off due to half the maximum projected area in elevation of the higher walls above the lowest, calculated at the selected design rate of rainfall, should be added.

Sloping and Vertical Surfaces

SLOPING ROOFS FREELY EXPOSED

Rain gauges do not take into account the driving effect of wind current with rain. Allowance for this is not required when designing drainage for horizontal surfaces or other surfaces protected from the wind, but should be considered where sloping or vertical surfaces occur that are freely exposed to the wind. Relevant information is sparse but at times of peak rainfall an angle of descent of wind-driven rain of one unit horizontal for each two units of descent should be allowed. Allowance for this may be made by adding to the run-off calculated from the plan area of that part of the roof draining to the gutter, an additional run-off from an area equal to half the maximum elevation area of that part of the roof. Use $a = b + \dfrac{c}{2}$

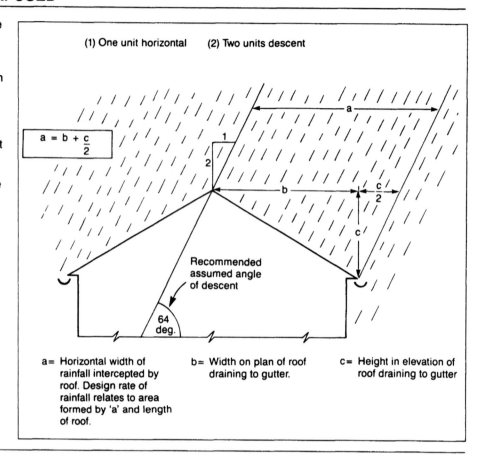

(1) One unit horizontal   (2) Two units descent

$$a = b + \frac{c}{2}$$

Recommended assumed angle of descent

64 deg.

a= Horizontal width of rainfall intercepted by roof. Design rate of rainfall relates to area formed by 'a' and length of roof.

b= Width on plan of roof draining to gutter.

c= Height in elevation of roof draining to gutter

VERTICAL SURFACES FREELY EXPOSED

Walls and other vertical surfaces unprotected by nearby structures of similar height will produce rainwater run-off. Where a wall drains onto a lower roof, provision should be made in the drainage of the latter for the additional flow. If only one wall is involved, allow for a catchment of half the actual vertical area at the full design rate of rainfall. No allowance should be made for absorption. If an angle or bay is formed by two or more walls, the direction from which these walls, considered as one unit, present the greatest projected area in elevation should be found and half that area taken at the design rate of rainfall to determine the run-off.

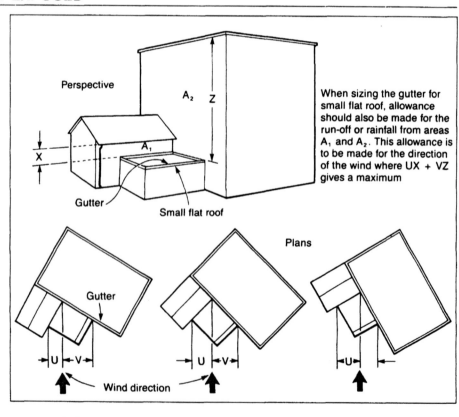

Perspective

A_2 Z

A_1

X

Gutter

Small flat roof

When sizing the gutter for small flat roof, allowance should also be made for the run-off or rainfall from areas A_1 and A_2. This allowance is to be made for the direction of the wind where $UX + VZ$ gives a maximum

Plans

Gutter

U   V

U   V

U

Wind direction

Sanitation: Roof Drainage

EXAMPLE INCORPORATING SINGLE VERTICAL SURFACE

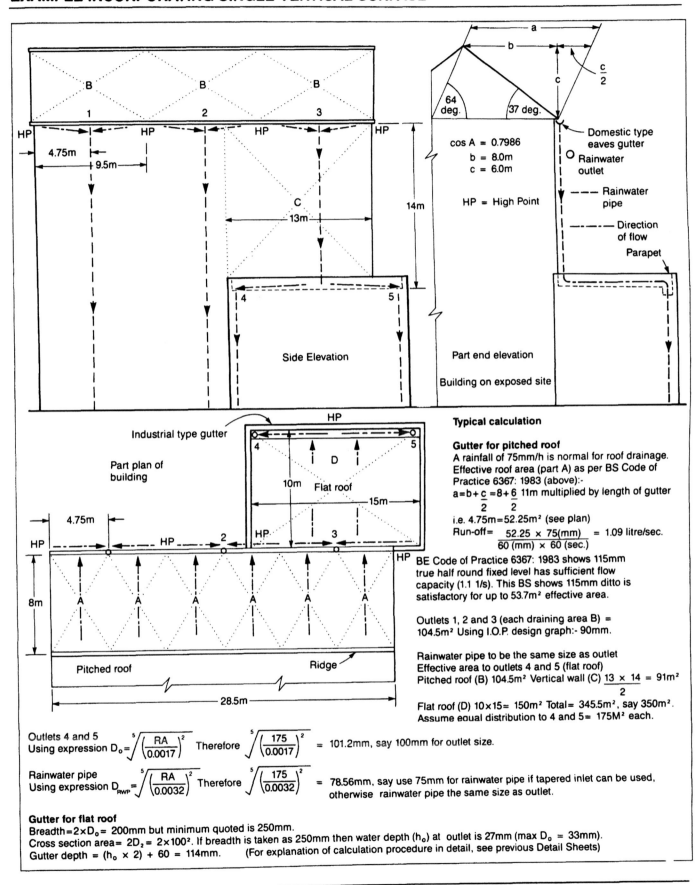

Typical calculation

Gutter for pitched roof
A rainfall of 75mm/h is normal for roof drainage.
Effective roof area (part A) as per BS Code of
Practice 6367: 1983 (above):-
$a = b + \dfrac{c}{2} = 8 + \dfrac{6}{2}$ 11m multiplied by length of gutter
i.e. 4.75m = 52.25m² (see plan)
Run-off $= \dfrac{52.25 \times 75(mm)}{60\,(mm) \times 60\,(sec.)}$ = 1.09 litre/sec.

BE Code of Practice 6367: 1983 shows 115mm
true half round fixed level has sufficient flow
capacity (1.1 1/s). This BS shows 115mm ditto is
satisfactory for up to 53.7m² effective area.

Outlets 1, 2 and 3 (each draining area B) =
104.5m² Using I.O.P. design graph:- 90mm.

Rainwater pipe to be the same size as outlet
Effective area to outlets 4 and 5 (flat roof)
Pitched roof (B) 104.5m² Vertical wall (C) $\dfrac{13 \times 14}{2}$ = 91m²

Flat roof (D) 10×15= 150m² Total= 345.5m², say 350m².
Assume equal distribution to 4 and 5= 175M² each.

Outlets 4 and 5
Using expression $D_o = \sqrt[5]{\left(\dfrac{RA}{0.0017}\right)^2}$ Therefore $\sqrt[5]{\left(\dfrac{175}{0.0017}\right)^2}$ = 101.2mm, say 100mm for outlet size.

Rainwater pipe
Using expression $D_{RWP} = \sqrt[5]{\left(\dfrac{RA}{0.0032}\right)^2}$ Therefore $\sqrt[5]{\left(\dfrac{175}{0.0032}\right)^2}$ = 78.56mm, say use 75mm for rainwater pipe if tapered inlet can be used, otherwise rainwater pipe the same size as outlet.

Gutter for flat roof
Breadth=2×D_o= 200mm but minimum quoted is 250mm.
Cross section area= $2D_2$ = 2×100². If breadth is taken as 250mm then water depth (h_o) at outlet is 27mm (max D_o = 33mm).
Gutter depth = (h_o × 2) + 60 = 114mm.   (For explanation of calculation procedure in detail, see previous Detail Sheets)

Material	BS	BS Title	Notes	BS. Sizes
Aluminium ● CO-EFFICIENT OF LINEAR EXPANSION Cast: 0.000026 deg. C Sheet: 0.000023 deg. C	2997	Aluminium rainwater goods	Aluminium rainwater goods are available in both cast and sheet material. The cast goods are of thicker section and consequently are stronger, more rigid and will withstand rough handling. Aluminium weathers well and generally need not be painted except in heavily polluted atmospheres (industry or marine), or for decorative reasons. Galvanic action may take place when this metal is in contact with other materials e.g. cast iron and steel. No copper or copper alloy components should be used in conjunction with Al and run-off from a copper roof onto Al must be avoided. Al is attacked when in contact with, or the run-off from concrete, mortar or plaster. Protect with bitumen.	See gutters/pipe table below

Aluminium — Gutters

Sheet		Sheet	
(HR/REC)	(OG)	(HR)	(OG)
75 mm	—	—	—
89 mm	—	—	—
100 mm	100	100	100
114 mm	114	114	114
125 mm	125	125	125
150 mm	150	150	—

Aluminium — Pipe (Sheet or Cast — Circular or Square)

50 mm	63 mm
75 mm	100 mm

Material	BS	BS Title	Notes
Asbestos cement CO-EFFICIENT OF LINEAR EXPANSION 0.000008 deg. C	569	Asbestos cement rainwater goods	Made of asbestos fibres and Portland cement, this material is strong and rigid but is liable to damage if subjected to rough handling. Gutters and fittings are normally supplied in natural grey but can be supplied bitumen coated or with a 'Colourglaze' finish. AC is durable, low in initial cost, non-combustible and has life of at least 40 years except in unfavourable conditions. Mould growth can be treated with fungicide.

Asbestos cement — BS. Sizes

Gutter		Pipe
(HR)	(OG)	(Circ)
75 mm	100 mm	50 mm
100 mm	114 mm	63 mm
114 mm	125 mm	75 mm
125 mm	150 mm	100 mm
150 mm	200 mm	150 mm
200 mm	—	—

Material	BS	BS Title	Notes
Cast iron CO-EFFICIENT OF LINEAR EXPANSION 0.0000106 deg. C	460	Cast iron rainwater goods	Cast iron is strong and rigid with a large range of fittings available for both industrial and domestic use. This material needs protection all its life. Light sections are usually supplied primed and heavier sections coated with bitumen solution. External pipes should be fitted with stand-off ears, bobbins or holderbats so that subsequent painting can be continuous to avoid the rapid corrosion of small unpainted areas.

Cast iron — BS. Sizes

Gutter	Pipe
(HR)	(Circ)
75 mm	50 mm
100 mm	63 mm
114 mm	75 mm
125 mm	100 mm
150 mm	125 mm
—	150 mm

Material	BS	BS Title	Notes
Copper ● CO-EFFICIENT OF LINEAR EXPANSION 0.000017 deg. C	1431	Wrought copper and wrought zinc rainwater goods	Copper is a light and durable material for rainwater goods; is not liable to corrosion and therefore need not be painted and requires little maintenance. In the course of time it develops green patina which forms a protection against corrosion by the effect of sulphurous gases in the atmosphere. This insoluble layer remains virtually unchanged with pleasant appearance. For long strip copper gutters, use ¼H or ½H temper, rectangular sections and beaded.

Copper — BS. Sizes

Gutter	Pipe
(HR/RECT/OG)	(Circ)
75 mm	50 mm
100 mm	63 mm
114 mm	75 mm
125 mm	100 mm
	(Rect)
	63 × 50
	89 × 75

Sanitation: Roof Drainage

Material	BS	BS Title	Notes	BS. Sizes
Lead ● CO-EFFICIENT OF LINEAR EXPANSION 0.000029 deg. C	1178	Milled lead sheet and strip for building purposes	Lead is a comparatively soft material requiring protection from possible mechanical damage. It is not generally liable to corrosion because the tarnishing caused by the oxygen, CO_2 and water vapour is a film which acts as a protective coating. Lead does not require painting. Water running over porous cement products can attack lead and cause build-up of carbonate scale. Lead rainwater pipes etc.. used as ornamental features.	Formed to suit requirements
Mild steel ● CO-EFFICIENT OF LINEAR EXPANSION 0.000011 deg. C	1091	Pressed steel gutters, rainwater pipes, fittings and accessories	Mild steel is strong and rigid but needs adequate protection against atmospheric corrosion. Protection can be achieved by coloured enamelling or galvanising. The thicker and complete the coatings, the longer the useful life, but care must be taken against damage e.g. chipping.	Gutter (OG and HR): 75 mm 114 mm; 89 mm 125 mm; 100 mm 150 mm — Pipe (Circ): 50 mm; 63 mm; 75 mm; 100 mm
Unplasticied polyvinyl chloride (uPVC) CO-EFFICIENT OF LINEAR EXPANSION 0.00006 deg. C	4576 Part 1	uPVC rainwater goods. Half round gutters and circular pipe	uPVC is a very light material, easy to assemble, precision made, free from rust, rot or corrosion, virtually eliminates maintenance and is chemical-resistant. Normal softening point:- above 80 deg. C. Abuse must be avoided at low temperatures because of its tendency to become brittle, otherwise it is a tough, flexible material with resistance to light degradation. It is supplied in several colours and although need not be painted, if other colours required, use only top coat paint. It has a relatively high co-efficient of thermal expansion and this must be positively allowed for. Normally factory made 'systems' are used.	Gutter (HR): 100 mm; 114 mm; 125 mm; 150 mm — Pipe (Circ): 63 mm; 2½ in; 75 mm; — Pipes to BS.4514 (Soil and Waste) can be used e.g. 75 mm, 100 mm and 150 mm
Zinc ● CO-EFFICIENT OF LINEAR EXPANSION 0.000026 deg. C Miscellaneous	1431	As for copper	Zinc is light and, because it is susceptible to atmospheric corrosion, the thicker the gauge selected, the longer the life span. This metal forms its own protective coating (a basic carbonate) after short exposure to the elements, and may have a life of up to 40 years. Detailed specifications may be different for internal and external use and care should be taken to select a pipework system to withstand the maximum hydraulic head which could occur should a blockage take place at the lowest point (see Detail 73 for warning pipes). Regarding action between dissimilar metals, examples from electro-chemical series:- Copper + 0.345V normal electrode potential, Lead — 0.125, Iron — 0.44, Zinc — 0.76, Al — 1.67.	See Copper for sizes

Al — aluminium. AC = asbestos cement. ¼H = quarter hard temper. ½H = half hard temper.
Note: Pitch fibre pipe with polypropylene fittings can be used internally when fixed in accordance with CP.304.
● Suitable for non-standard sections

Fixing of Domestic Eaves Gutters and Rainwater Pipes

GUTTERS

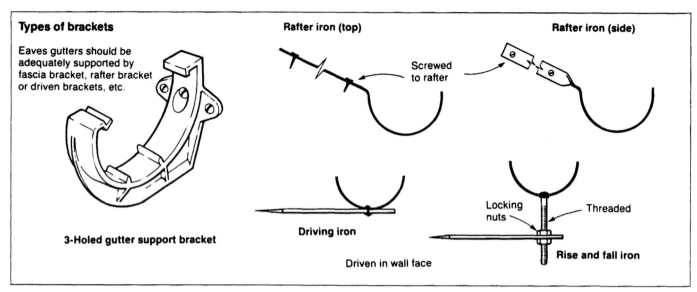

Types of brackets

Eaves gutters should be adequately supported by fascia bracket, rafter bracket or driven brackets, etc.

3-Holed gutter support bracket

Rafter iron (top)

Screwed to rafter

Driving iron

Driven in wall face

Rafter iron (side)

Locking nuts — Threaded

Rise and fall iron

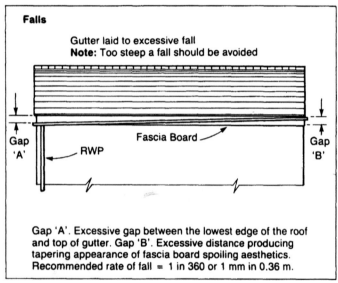

Falls

Gutter laid to excessive fall
Note: Too steep a fall should be avoided

Gap 'A'
RWP
Fascia Board
Gap 'B'

Gap 'A'. Excessive gap between the lowest edge of the roof and top of gutter. Gap 'B'. Excessive distance producing tapering appearance of fascia board spoiling aesthetics. Recommended rate of fall = 1 in 360 or 1 mm in 0.36 m.

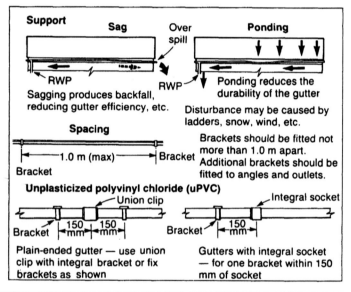

Support

Sag — Over spill — **Ponding**

RWP
Sagging produces backfall, reducing gutter efficiency, etc.

RWP — Ponding reduces the durability of the gutter

Disturbance may be caused by ladders, snow, wind, etc.

Spacing

Bracket — 1.0 m (max) — Bracket

Brackets should be fitted not more than 1.0 m apart. Additional brackets should be fitted to angles and outlets.

Unplasticized polyvinyl chloride (uPVC)

Union clip
Bracket — 150 mm — 150 mm

Integral socket
Bracket — 150 mm

Plain-ended gutter — use union clip with integral bracket or fix brackets as shown

Gutters with integral socket — for one bracket within 150 mm of socket

POSITIONING

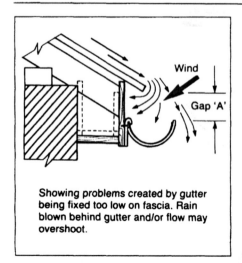

Wind
Gap 'A'

Showing problems created by gutter being fixed too low on fascia. Rain blown behind gutter and/or flow may overshoot.

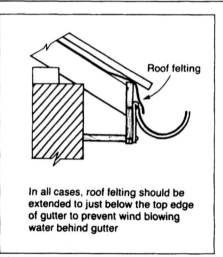

Roof felting

In all cases, roof felting should be extended to just below the top edge of gutter to prevent wind blowing water behind gutter

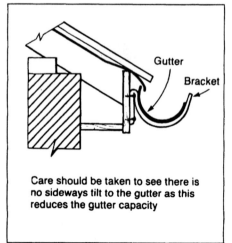

Gutter
Bracket

Care should be taken to see there is no sideways tilt to the gutter as this reduces the gutter capacity

Sanitation: Roof Drainage

FIXING

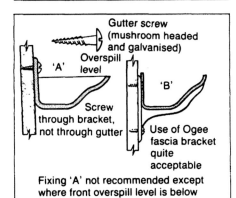

Screws grip better if they penetrate

Foot of rafter

Fixing screws should be stout brass or zinc plated 25 mm long × 5 mm wood screws

Gutter screw (mushroom headed and galvanised)

Overspill level

'A'

'B'

Screw through bracket, not through gutter

Use of Ogee fascia bracket quite acceptable

Fixing 'A' not recommended except where front overspill level is below level of fixing screw

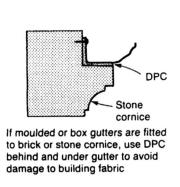

DPC

Stone cornice

If moulded or box gutters are fitted to brick or stone cornice, use DPC behind and under gutter to avoid damage to building fabric

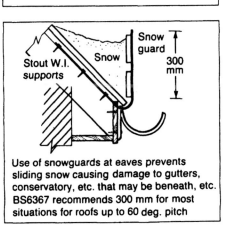

Snow guard

Snow

300 mm

Stout W.I. supports

Use of snowguards at eaves prevents sliding snow causing damage to gutters, conservatory, etc. that may be beneath, etc. BS6367 recommends 300 mm for most situations for roofs up to 60 deg. pitch

RAINWATER PIPES

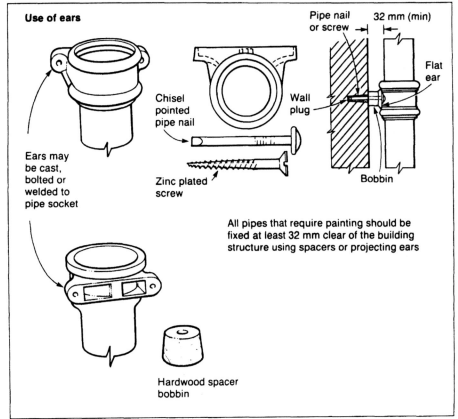

Use of ears

Ears may be cast, bolted or welded to pipe socket

Chisel pointed pipe nail

Zinc plated screw

Pipe nail or screw

32 mm (min)

Wall plug

Flat ear

Bobbin

All pipes that require painting should be fixed at least 32 mm clear of the building structure using spacers or projecting ears

Hardwood spacer bobbin

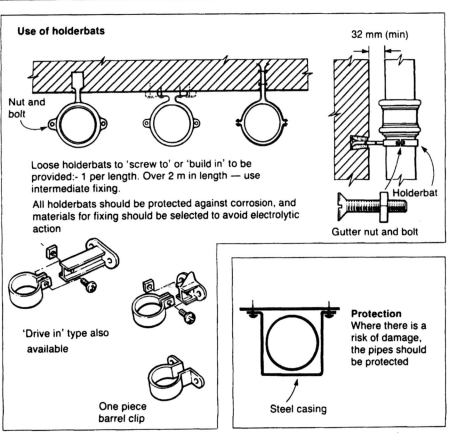

Use of holderbats

Nut and bolt

Loose holderbats to 'screw to' or 'build in' to be provided:- 1 per length. Over 2 m in length — use intermediate fixing.

All holderbats should be protected against corrosion, and materials for fixing should be selected to avoid electrolytic action

'Drive in' type also available

One piece barrel clip

32 mm (min)

Holderbat

Gutter nut and bolt

Protection
Where there is a risk of damage, the pipes should be protected

Steel casing

RWP = Rainwater pipe
Bkt. = Bracket
WI = Wrought iron

Jointing of Domestic Eaves Gutters

GUTTERS

Materials
*1. Red lead putty
*2. Bituminous or other mastic compound
*3. Special mastic specified by the gutter manufacturer
*4. Pre-formed neoprene or other plastic strips 'mastics'

Notes on Mastics When using any form of mastic it is essential to ensure an even spread of the mastic over the whole surface of the jointing socket. The action of tightening the fixings should cause a certain amount of mastic to squeeze out of the gutter joint. This should always be cleaned off, leaving a neat joint externally and with no mastic protruding internally to obstruct flow.

Notes on Strips Certain neoprene or other plastic strips can be laid directly in the socket before fixing (some strips have a pressure-sensitive adhesive with paper backing on one face). Others are supplied with the strip already fixed in position in the socket. When tightening the joint, care should be taken not to over-compress the material. Use manufacturers' advice.

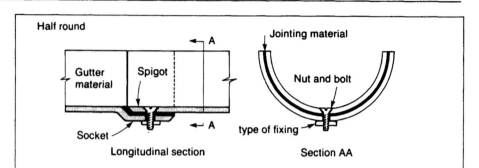

Half round · Gutter material · Spigot · Socket · Longitudinal section · A—A · Jointing material · Nut and bolt · type of fixing · Section AA

Joint Fixing
1. Protected steel or aluminium gutter bolts
2. Patent clips forming part of gutter socket.
3. Lapping and soldering

Jointing. The three variables are gutter material, jointing material, and type of fixing and it is important that the correct combination of materials is always used. The advice of manufacturers on the methods of jointing should be followed.

Gutter materials	Jointing Materials	Joint Fixing	Notes
Cast iron and steel	Red lead putty, mastic compound or pre-formed strips	Protected steel nuts and bolts	Paint spigot and socket and cut ends of non-ferrous gutters before jointing to prevent corrosion
Asbestos cement	Special mastic (refer to manufacturers) or pre-formed strips	Protected steel nuts and bolts	Extra length sockets to have length of tarred yarn inserted at end of socket for full width if mastic is used
Aluminium	Mastic compound or pre-formed strips	Aluminium nuts, bolts and washers	Electrolytic action should be guarded against. Avoid use of dissimilar metals
Unplasticized polyvinyl chloride	Pre-formed strips (these are very often pre-fixed in the socket)	The joint clips together as patented by manufacturer	Allowance should be made in each joint for thermal movement of the gutter
Copper and zinc	Lapping and soldering for a minimum of 38 mm in the direction of the flow	Solder	Pre-tin copper and pre-flux zinc before jointing, solder along the full girth of gutter

THERMAL MOVEMENT

Gutters. Supports and fixings to gutters should allow for thermal movement to take place and, in addition, expansion joints may be necessary. Spacing of expansion joints depends upon the flexibility of the jointing material used, the method of jointing and supporting, and the co-efficient of expansion of the material of which the gutter is made. Except where the methods of jointing and fixing provide adequate allowance, very long lengths may have to be divided into suitable sections. The allowance for expansion may then be by means of a gap between sections, suitably weathered. Where the ends of gutters abut a structure (e.g. gutters fixed between brick walls), a suitably weathered gap should be left between gutter end and structure. Structural and gutter expansion joints should coincide.

Rainwater pipes. The type of jointing used for rainwater pipes should allow sufficient movement for thermal expansion to take place without leakage, distortion and displacement of fittings. Particular care is necessary when rainwater pipes of long length are used.

Thermal movement in upvc systems. The coefficient of expansion of upvc is 5×10^{-5} per deg. C giving a movement of about 0.5 mm per m length per 10 deg. C change in temperature. This thermal movement should be allowed for, and is accommodated for within the fixing brackets to the fascia. Gutter and pipe fittings are designed to allow for expansion and contraction.

Sanitation: Roof Drainage

RAINWATER PIPES

External fixing. Joints on vertical spigot and socket rainwater pipes are generally left unsealed. Metal pipes should be wedged to prevent rattling. For neat appearance for joints below eye level, they may be wedged to centralise and the joint partially filled with red lead putty and neatly 'trimmed'.

Internal rainwater pipe. Must be jointed as for soil pipes.

Horizontal runs. External and internal joints must be properly sealed and made watertight.

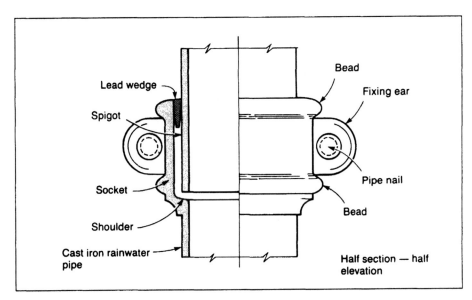

Lead wedge
Spigot
Socket
Shoulder
Cast iron rainwater pipe
Bead
Fixing ear
Pipe nail
Bead
Half section — half elevation

Pipe materials	Jointing materials	Notes
Metal pipes	Red lead putty or mastic compound	For medium and heavy cast iron pipes, a ring of spun yarn should first be introduced to prevent jointing material entering the inside of the pipe
Asbestos cement pipes	Suitable jointing compound as recommended by manufacturers	Stiff mortar composed of one part of Portland cement and two parts of sand (by volume) may be used as a jointing material
Unplasticized polyvinyl chloride	Many patent joints are marketed (refer to manufacturers literature). Loose sockets are provided for fitting to plain-ended pipes	A gap should be left between the spigot end of the pipe and the shoulder in the socket in order to allow for thermal movement. Systems with 'sealed' joints are available for use where this type of joint is required.

Drainage of Paved Areas

IRREGULAR SHAPES

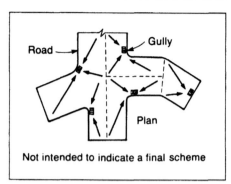

Not intended to indicate a final scheme

Irregular shape of areas between buildings will often determine the number of outlets rather than the permissible area which can be drained to one outlet.

EDGE FINISH

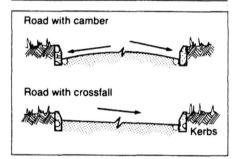

Paved areas subjected to vehicular traffic may involve the additional construction of kerbed boundaries to carriageways and verges.

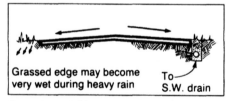

Subsoil and the absence of heavy loading permitting, the edges of areas may finish against and level with the ground. This should not be done where heavy traffic comes to edge of paved area.

INTERCEPTORS

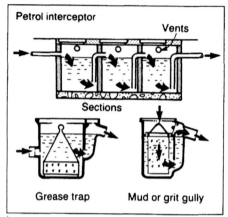

Materials which may be a source of pollution, danger of silting-up etc. should be excluded from the s.w. drainage system.

SITING OF GULLIES

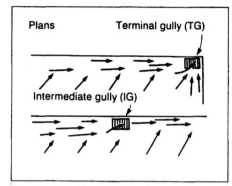

Terminal Gullies Sited at lower points; should be generously sized as they are more likely to cause flooding than IGs.

Intermediate Gullies Sited to avoid build-up of excessive flows in channels falling towards terminal gullies. Some flow may leap over or skirt round the gully.

CHANNEL CUT-OFFS AND ROAD ENTRANCES

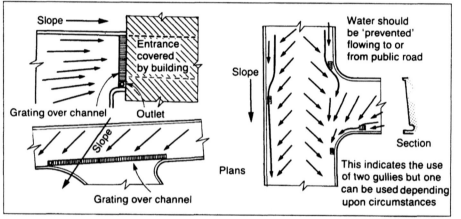

Where an area is continued into buildings or below paved area level, such as underground garages, water should be prevented from entering by using such devices as channels provided with gratings. Draining and positioning of gullies at entrances should be arranged to reduce water flowing across the entrance.

REVERSE FALL

The provision of a reverse fall strip, falling to some type of channel, is strongly recommended.

Water should not be drained to concentrate along the side of a building. A reverse fall on a narrow strip of yard should be used to minimise this problem.

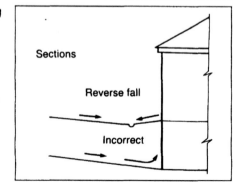

LARGE AREAS

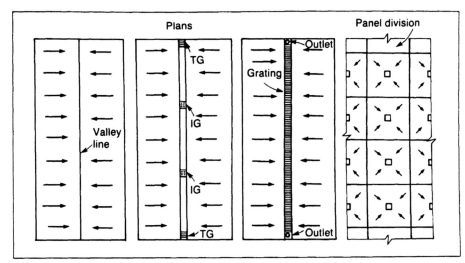

Plans | Panel division

On large areas it may be more convenient to drain to a central channel rather than to divide the area into panels, each with a number of central collecting points. The central channel may be drained to terminal gullies with intermediate gullies if necessary. Other arrangements could include a simple valley line, channel (open or 'grated'), etc.

GRADIENTS AND ECONOMIC SPACING OF GRATINGS

	Gradients		
	Access Roads	Paved Areas	Footpaths
Longitudinal gradient or fall	1 in 15 (max)*		
Camber or crossfall	1 in 40 normal	1 in 60 (min)	1 in 30 to 1 in 40 (min) (max)
Kerb channels (no channel blocks)	1 in 120 (min)	1 in 120 (min)	
Kerb channels (with channel blocks or high class surfacing)	1 in 200 (min)	1 in 200 (min)	
Super elevation Road curves not exceeding 100 m rad.	1 in 25 (max)		

Table 8 (BS6367) of gradients for paved areas.

*The first 10 m of an access road from its junction with major road or public highway should have a gradient of not more than 1 in 30.

A method for determining the economic spacing of gratings has been developed by John Hopkins University and a summary of this is given in Appendix C BS6367:1983.

GULLY POTS

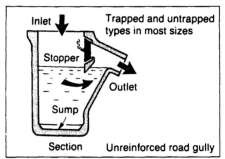

Section   Unreinforced road gully

Gullies should normally be provided with grit 'interceptors' of adequate size (see mud or grit gully above) depending upon the use of the area, the type of surface and the frequency of cleaning. Where a gully discharges into a partially separate system or combined, a water-sealed trap is required and such a trap is normally part of the gully 'pot'. Traps may be omitted from individual gullies which discharge into a drain used solely for the purpose of conveying rainwater from a roof, provided that such a drain is fitted with a suitable trap before it discharges into a foul water drain. L.A. may require traps to gullies discharging into surface water sewers, soakaways or watercourses.

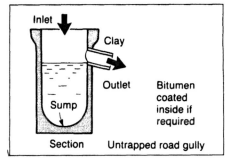

Section   Untrapped road gully

Composition of refuse

Refuse is not homogenous and does not have consistent characteristics and domestic refuse has been described as a heterogeneous compound of organic and inorganic wastes. The inorganic (mostly carbonaceous) includes paper, cardboard, textiles, ash, vegetable waste; and the inorganic includes broken crockery, glass, ferrous and non-ferrous metals, stones, etc. If we accept that domestic refuse comprises everything discarded after use, it will include not only cinders, ashes, tea leaves, fruit, vegetable and animal wastes, paper, cardboard, plastics, kitchen equipment, cans, glass and crockery, but also bedsteads, bedding, broken or discarded furniture, worn-out motor cars, etc. Much of the latter can be described as bulky waste (i.e. all articles too large to be contained in a dustbin), and accounts for ¾ percent (by weight) of the total waste handled by local authority and 6/8 per cent by volume (does not include cars). Amounts vary due to local and seasonal variations such as coal mining towns, seaside resorts, good-class residential areas, all-electric flats, premises with garbage grinders. Allowances should be made, if cases are known, that the occupants will differ e.g. old people's flats or maisonettes. Also allowance should always be made to provide sufficient storage at winter (and perhaps other) bank holiday periods.

Trend

The quantity produced rises with our standard of living and its composition is gradually changing; more abundant packaging and fewer solid fuel burning appliances have resulted in an increased volume of lower density waste and this trend is likely to continue. At the same time land is becoming scarcer and space in which to tip refuse is increasingly hard to find.

Definition of containers

Binette

Small portable refuse container with lid for use within a dwelling, eventually to be emptied.

Individual refuse container (dustbin or sack)

Container with a capacity not exceeding $0.11m^3$ (4 cu.ft.) in which refuse is stored awaiting collection.

Refuse storage container

Movable refuse container, capacity not exceeding $1m^3$ (1¼ cu.yd) storing refuse for collection.

Bulk container

Movable container, capacity up to $9.17m^3$ (12 cu.yd).

Estimating yield

The storage capacity provided must be related to the frequency of collections, the nature and volume of refuse. The amount per dwelling in 1969 was around 13kg per week, based on an average of 2.8 persons per dwelling (national average) or approximately $0.1m^3$ per week. Where a $0.07m^3$ bin is used, refuse must be compressed in order to get it into this dustbin and many users have a larger bin ($0.09m^3$) or more than one bin.

Some relevant facts

Solid fuel consumption fell from 39.6m tonnes in 1955 to 29.5m tonnes in 1965. Gas and oil consumption doubled. 14.2m tonnes of refuse = volume of $68.8mm^3$. Cost in 62/63 was £40m/annum (average £2.50 per property), and £45m in 1965/66 (average £2.80 per property).

Notes on Legal Requirements

Introduction
The principal Act is the Public Health Act 1936. There is no legal definition of refuse but the Oxford dictionary states — 'that which is cast aside as worthless; rubbish or worthless matter of any kind; the rubbishy part of anything.' A dustbin is legally defined (S.343) '. . .a movable receptacle for the deposit of ashes or refuse.' Paper sacks with their holders appear to be within this definition, so they are governed by the legal provisions applying to dustbins. The Act distinguishes between (a) house refuse, (b) trade refuse and (c) refuse which the Local Authority 'are under no obligation to remove' but gives no definitions. There seems to be general agreement that the most important consideration is the character of the refuse and that house refuse is the sort of refuse which arises from the ordinary domestic occupation of a house. Hotels, for example, can also produce 'house refuse'.

Power and duties of Local Authorities
Councils of the Inner London Boroughs have a statutory duty to collect house refuse. Elsewhere, the local Public Health authorities may, and if required by the Minister must, undertake the removal of house refuse from the whole or part of their district — without charge so far as their undertaking extends. Except in Inner London, a local Public Health authority can rescind their undertaking to remove house refuse, but, if it is being carried out under a Minister's resolution, his consent is required. Refuse collecting authorities may make byelaws (S.72(3)):-
(a) imposing duties on occupiers to facilitate collection; (b) requiring the use of dustbins provided by the authority; (c) prohibiting the deposit of liquids in bins; (d) *controlling the deposit of refuse in ashpits or bins*; (e) prohibiting the removal of the refuse except by the Council.

Trade refuse
The Inner London Boroughs have a statutory duty to collect any trade refuse, at the request of the occupier of the premises. Elsewhere, a Local Authority may undertake the removal (similarly on request) of trade refuse. There is no provision whereby the Minister can require them to collect. All Local Authorities collecting trade refuse must make reasonable charges; they cannot legally make a general collection of trade refuse without charge. Except for the Inner London Boroughs, they have an implied power to rescind a resolution to collect trade refuse. Disputes on what is trade refuse or what is a reasonable charge for removing it may be determined by a court of summary jurisdiction.

Storage of refuse
A Local Authority can require the owner or occupier of premises from which they have undertaken the removal of house refuse to provide one or more covered dustbins of such material, size and construction as the Local Authority may approve. In the event of non-compliance, or if the bins are not kept in good condition, the Local Authority can provide bins themselves and recover the cost reasonably incurred. The person in default may also be fined; similar powers in Inner London deal with dustbins for trade refuse. Instead of requiring owners and occupiers of buildings to provide and maintain dustbins for house refuse (or for trade refuse in Inner London) the Local Authority may undertake to provide them. For each dustbin they supply, they may make an annual charge up to a maximum fixed by the Minister, and they may recover the charge as part of the general rate for the premises. There is no provision in the Act expressly authorising Local Authority to finance the supply of municipal bins from the general rate, but it could be so interpreted, and many Local Authorities have acted upon this interpretation instead of making a charge for each bin supplied. Some Local Authorities have taken local powers to put the matter beyond doubt.

Sanitation: Refuse From Buildings

Frequency of collection
The general law does not specify any minimum frequency of collection. The Minister has no power to prescribe the frequency where a Local Authority has voluntarily undertaken this service. In the area of the former County of London, byelaws made by the LCC (which are still in force) require the London Borough Councils to collect house refuse at least once a week, unless unavoidably prevented.

Access to premises
Local Authorities (other than Inner London Boroughs, which have separate legislation) are required to reject building plans for the erection or extension of a house unless a satisfactory access to a street for the purpose of refuse removal is to be provided. It is unlawful to close such access without consent of Local Authority.

Reorganisation of Local Authorities
Since 1st April 1974, new district authorities in England are responsible for the collection of house and trade refuse and the new counties are responsible for its disposal. In Wales, new district councils are responsible for collection and disposal. The Government is at present considering proposals for the introduction of legislation which will extend and develop existing statutory controls over waste collection and disposal. It is proposed to define household, trade and industrial wastes according to the kind of premises from which the waste originated and to redefine the powers and duties of Local Authority in respect of each category of waste. In London, the Inner Boroughs collect and dispose.

Domestic Systems in Outline

WATERBORNE TO SITE STORAGE (GARCHEY)

Designed for blocks of flats. Specially designed sink in each flat. Refuse is flushed into bowl and retained until a 'valve' is operated, allowing water and refuse to discharge into vertical stack pipe.

Suction to emptying vehicle (compresses to 20 per cent to 30 per cent of original volume)

Vehicle contents taken to tip or incinerator

Surplus water returned from vehicle

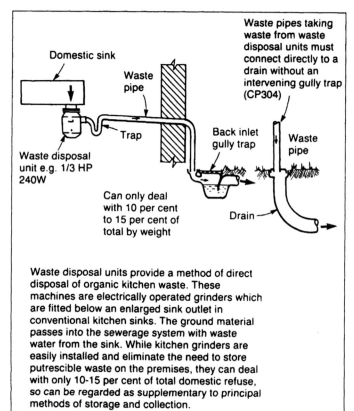

Waste pipe for liquid

Domestic sink in kitchen

Vent pipes

Branch to stack

Special bowl unit

Plumbers' waste pipe

'Garchey' refuse stack pipe

Underground collection chamber

Overflow

Sewer

Can deal with 50 to 60 per cent of total volume (by weight)

System has advantage of allowing housewife to get rid of a proportion of refuse directly within the dwelling. Will accept bottles and tins but does not readily handle large amounts of paper.

WATERBORNE TO SEWER

Domestic sink

Waste pipe

Waste pipes taking waste from waste disposal units must connect directly to a drain without an intervening gully trap (CP304)

Trap

Waste disposal unit e.g. 1/3 HP 240W

Back inlet gully trap

Waste pipe

Can only deal with 10 per cent to 15 per cent of total by weight

Drain

Waste disposal units provide a method of direct disposal of organic kitchen waste. These machines are electrically operated grinders which are fitted below an enlarged sink outlet in conventional kitchen sinks. The ground material passes into the sewerage system with waste water from the sink. While kitchen grinders are easily installed and eliminate the need to store putrescible waste on the premises, they can deal with only 10-15 per cent of total domestic refuse, so can be regarded as supplementary to principal methods of storage and collection.

DRY CHUTE TO CONTAINER

The chute(s) is a vertical pipe running from top to bottom of the building and, basically, on each floor there is a hopper which opens into it. The chute(s) discharges into storage containers in specially constructed chambers at or near to ground level. Chutes need to be well designed and planned to prevent fire risk and to minimise blockages and the possibility of dust, smell and noise inside the building. The chute is convenient to the householder and will take most of the refuse, leaving little to be carried downstairs. The task of the L.A. is simplified because they can collect the refuse from a number of flats at one point. Regarded as a less unsatisfactory system for high density dwellings.

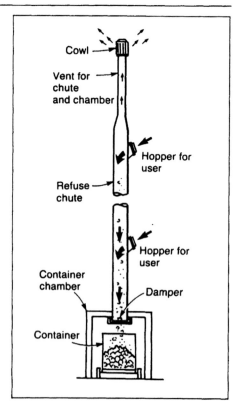

Cowl

Vent for chute and chamber

Refuse chute

Hopper for user

Hopper for user

Container chamber

Damper

Container

Sanitation: Refuse From Buildings

SACK COMPACTION

Containers will hold more if the refuse is compacted and refuse compressors are available for use under chutes in blocks of flats and large buildings. Up to 10 'sacks' are mounted on a turntable installed in the chamber, and a cut-off plate (not shown) is provided to seal the chute whilst compaction is taking place by means of a ram. After repeated filling and compression, the refuse in the sack reaches a predetermined level and turntable rotates to bring an empty sack under the chute.

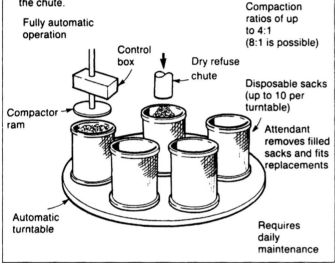

Fully automatic operation

Compaction ratios of up to 4:1 (8:1 is possible)

Control box

Dry refuse chute

Disposable sacks (up to 10 per turntable)

Compactor ram

Attendant removes filled sacks and fits replacements

Automatic turntable

Requires daily maintenance

PNEUMATIC CONVEYING SYSTEMS

This system requires the use of conventional refuse chutes or central loading points at ground floor level. The lower end of the chute terminates in a valve chamber, where the chute is connected by 'lead-in' pipes to horizontal transporter pipes of approx. 525 mm diameter, laid below ground. Refuse falling down the chute, rests on a flat 'valve' disc to be stored and creates a seal when vacuum is being created in the transporter pipes by extractor turbines. The disc 'valves' are operated at predetermined intervals (2 to 4 times a day) and the refuse is carried along the pipe to silo. Travel distance of 2 km (+) is possible.

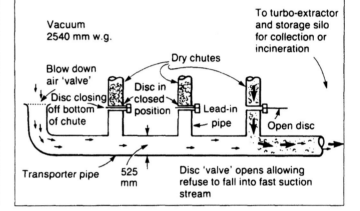

Vacuum 2540 mm w.g.

To turbo-extractor and storage silo for collection or incineration

Blow down air 'valve'

Dry chutes

Disc in closed position

Disc closing off bottom of chute

Lead-in pipe

Open disc

Transporter pipe

525 mm

Disc 'valve' opens allowing refuse to fall into fast suction stream

ON-SITE INCINERATION

This is a method whereby refuse is incinerated on site to reduce the initial volume by 7:1 to 10:1. On-site incinerators are chute fed, the refuse being automatically fed into the combustion chamber at predetermined times. Temperature probes are fitted to control the operation of the burners. Most on-site incinerators are designed to customers' requirements and due regard should be paid to any specific requirements of the Clean Air Acts 1956-68 relating to control of smoke, grit or dust emission. They require competent and regular supervision and management.

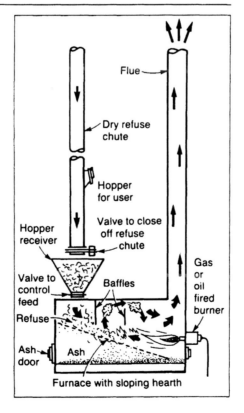

Flue

Dry refuse chute

Hopper for user

Valve to close off refuse chute

Hopper receiver

Gas or oil fired burner

Valve to control feed

Baffles

Refuse

Ash door

Ash

Furnace with sloping hearth

The following requirements are needed by the Building Regulations 1985 or BS.5906: 1980 Code of Practice for storage and on-site treatment of solid waste from buildings.

Refuse storage container chambers constructed in Buildings comprising more than one dwelling

— chamber walls, floor and roof to be made of suitable non-combustible materials.
— fire resistance of walls and floors that separate chamber from building to comply with Building Regulations (or one hour fire resistance whichever is the greater).
— the inner surfaces of the chamber to be impervious to moisture.
— flood laid to a fall towards a trapped gully (gully can be inside or immediately outside chamber).
— flush door in the external wall to afford ½ hour fire resistance and be sole means of access.
— refuse to be deposited in container by chute and/or hopper.
— where delivery is by hopper only directly into chamber (not by chute), then chamber to have flyproof vent as high as practicable or vent pipe shaft.

Refuse chutes in buildings comprising more than one dwelling

— to be constructed of suitable non-combustible material of such thickness and so arranged to prevent spread of fire within the chute or chamber, to any other part of the building.
— constructed so that inner surfaces are impervious to moisture.
— constructed to prevent lodgement of refuse within the chute.
— to be circular in cross-section with an internal diameter of not less than 375mm.
— to be filled with adequate means of access for inspection and cleansing.
— to be fitted with one or more hoppers for insertion of refuse and chute must be ventilated at top and bottom.
— to have shutter fitted at lower extremity capable of closing the outlet of the chute.

Pipes or shafts ventilating refuse storage container chambers or refuse chutes

— to be constructed of suitable non-combustible material of such thickness and so arranged to prevent spread of fire within the chute or chamber, to any other part of the building.
— to be not less than 17,000mm² in cross-sectional area (e.g. 150mm diameter if circular).
— chute to be carried upwards to such a height and so positioned as not to transmit foul air which may become prejudicial to health or nuisance, and the outlet protected against the entry of rain.

Hoppers for refuse storage container chambers or refuse chutes

— situate in place which is either freely ventilated or has adequate means of mechanical ventilation.
— to be constructed of suitable non-combustible material.
— so constructed and installed to discharge refuse into chute or container efficiently.
— door to be incapable of remaining in any position other than open or closed.
— to prevent as far as possible the emission of dust or foul air (whether open or closed).
— hoppers designed for refuse chutes must not project into the chute.
— no such hopper shall be situated within a dwelling.

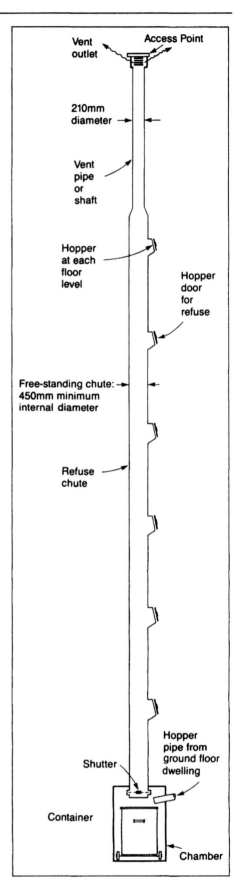

Vent outlet

Access Point

210mm diameter

Vent pipe or shaft

Hopper at each floor level

Hopper door for refuse

Free-standing chute: 450mm minimum internal diameter

Refuse chute

Hopper pipe from ground floor dwelling

Shutter

Container

Chamber

SOUND INSULATION

Sound insulation
Approved Documents E1, E2 and E3 of the Building Regulations 1985

Sound insulation of walls
— any wall which separates a habitable room in a dwelling from a refuse chute is required to have an average mass of not less than 1320 kg/m² (including plaster).
— in other cases, in a dwelling where the chute is not adjoining a habitable room, the average mass must be not less than 220 kg/m² (including plaster).

Section 1.4 refuse chutes relates to sound insulation for refuse chutes.

Fire spread
Approved Document B of the Building Regulations 1985.

A refuse chute falls into the meaning of a "Protected Shaft" as defined in paragraph 0.26 Introduction to Provisions — "any spaces connecting compartments need to be protected in a way that restricts fire spread". Appendix E to the Approved Document gives information that should be incorporated in its design:-

(a) The construction should form a complete barrier to fire between different compartments which the shaft connects;

(b) It has the appropriate fire resistance;

(c) Any beam or column which forms part of the enclosure shall be constructed with materials of limited combustibility, if its fire resistance is required to be 1 hour or more.

Information is also provided on its uses and ventilation.

Note: Attention is also drawn to the Public Health Act 1936 Section 55 Means of Access for Removal of Household Refuse. The Local Authority responsible for refuse collection should be consulted regarding their requirements respecting the capacity of refuse containers and frequency of collection, where chutes are proposed. The Local Authority may insist on the provision of refuse chutes in blocks of flats exceeding four storeys in height.

Design of Refuse Chambers

GENERAL NOTES

Facilities for the storage and collection of refuse should provide adequate storage combined with the maximum convenience for the occupier and refuse collector and the highest practicable standard of hygiene, amenity, safety from fire risk and smoke and sound insulation. The chamber should be located at vehicle access level (essential in the case of bulk containers) preferably away from the main entrance to the building. The position should be decided in conjunction with the design of chutes and the roadways and approaches to buildings. Separate enclosed accommodation at ground level in an accessible position should be provided for the storage of large and bulky articles so that the L.A. can make special collection arrangements. A minimum space of 10 m² is recommended or 0.03 m³ per person. Containers should be constructed in accordance with the BS1136 or BS3495. Where containers are proposed to be used for which no BS is available, these should be of a type approved by the L.A. CP306 defines the following: Refuse storage container:- A movable refuse container, with a capacity not exceeding 1 m³ in which refuse is stored awaiting collection. Bulk container:- A movable container, exceeding 1 m³ in capacity up to 9.17 m³ (12 yd³). (Larger sizes are available).

CONTAINERS

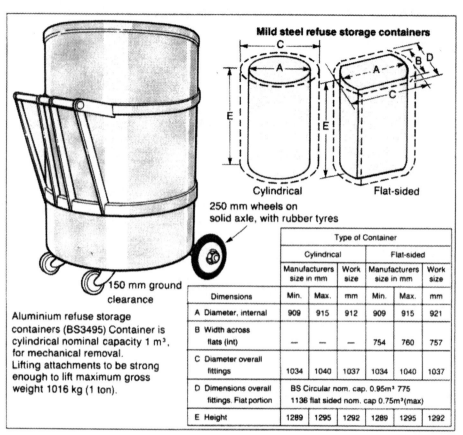

Mild steel refuse storage containers

Cylindrical   Flat-sided

250 mm wheels on solid axle, with rubber tyres

150 mm ground clearance

Aluminium refuse storage containers (BS3495) Container is cylindrical nominal capacity 1 m³, for mechanical removal.
Lifting attachments to be strong enough to lift maximum gross weight 1016 kg (1 ton).

Dimensions	Type of Container					
	Cylindrical			Flat-sided		
	Manufacturers size in mm		Work size	Manufacturers size in mm		Work size
	Min.	Max.	mm	Min.	Max.	mm
A Diameter, internal	909	915	912	909	915	921
B Width across flats (int)	—	—	—	754	760	757
C Diameter overall fittings	1034	1040	1037	1034	1040	1037
D Dimensions overall fittings. Flat portion	BS Circular nom. cap. 0.95m³ 775 1136 flat sided nom. cap 0.75m³(max)					
E Height	1289	1295	1292	1289	1295	1292

Bulk containers
Containers much larger than the standard 1 m³ size and made to different designs have been coming into use in the last few years for the storage and transport of rubble, spoil and refuse. Up to 9 m³ — for use with skip lift vehicle; up to 23 m³ — for use with lift hoist vehicles; up to 30 m³ — as trailers for articulated vehicles. These containers avoid the use of a multiplicity of smaller receptacles and require a minimum of manpower.
When used to service a chute in a tall block of flats, the additional headroom required in the chamber must be borne in mind and the special collection vehicle must be able to drive right up to them.

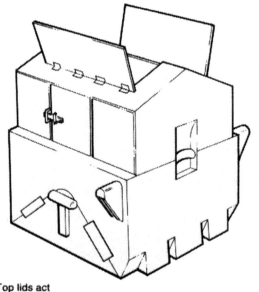

Top lids act as deflectors to prevent spillage

Size range:
4.6, 6, 7.6 and 9.2m³

Sanitation: Refuse From Buildings

REFUSE CHAMBERS

Minimum clear space 150 mm between containers and between containers and walls. Minimum general height of chamber for containers to BS1136 and 3495 should be 2m. For bulk containers, minimum head room 3m. Bottom edge of chute should finish minimum 25 mm below level of ceiling to form a drip; also maximum 225 mm between chute edge and top of container. Walls and roof to be non-combustible and impervious with fire resistance of 1 hour. The door of the chamber should be of steel or have a fire resistance of ½ hour, with a lock approved by L.A. The walls should be constructed of or lined with hard impervious material with a smooth finish suitable for washing down. Floor, minimum 100 mm thick, hard, impervious, smooth, with no steps or projections. Artificial lighting to be to BS889.

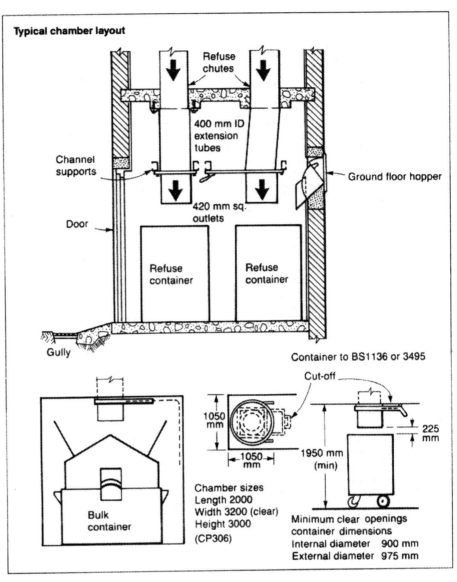

Typical chamber layout

Refuse chutes

400 mm ID extension tubes

Channel supports

Ground floor hopper

Door

420 mm sq. outlets

Refuse container

Refuse container

Gully

Container to BS1136 or 3495

Cut-off

1050 mm

1050 mm

225 mm

1950 mm (min)

Bulk container

Chamber sizes
Length 2000
Width 3200 (clear)
Height 3000
(CP306)

Minimum clear openings container dimensions
Internal diameter 900 mm
External diameter 975 mm

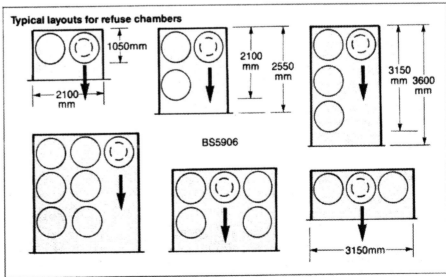

Typical layouts for refuse chambers

1050mm

2100 mm

2100 mm

2550 mm

3150 mm

3600 mm

BS5906

3150mm

Design of Refuse Chutes,

DESIGN NOTES. CHUTES

1. The number of and siting of chutes will be dependent upon the layout of the building, the systems of storage and collection and the means of access for the collecting vehicles and the volume of refuse.
2. The chute should be sited not more than 30 m measured horizontally from each dwelling.
3. In terms of building costs, it is more economical to provide space for additional storage beneath each chute, than to provide additional chutes. It is recommended that a minimum of two days' storage without attention by the caretaker be provided.
4. Normally chutes should be designed without bends or offsets throughout their length (except above the topmost hopper) and arranged so as to discharge centrally over the container. Where a change of direction is unavoidable, a bend — minimum 60 deg. to the horizontal — should be provided.
5. Chutes should be circular in section, minimum of 381 mm diameter, but preferably not less than 450 mm diameter.
6. In order to safeguard against the possible spread of fire, the chutes should be constructed of or enclosed by fire-resisting construction possessing minimum standard of 1 hour fire-resistance or such higher standard as may lawfully be required.
7. Noise nuisance can be considerable, therefore chutes should not be located adjacent to habitable rooms. If possible, isolate from dwelling.
8. They should be formed of (or lined with) a smooth impervious internal surface to BS1703.
   The pipes or linings should be flush-jointed to provide a continuous smooth surface and the number of joints kept to a minimum to avoid lodgement of refuse.
9. Chutes to have provision for access for removal of obstruction.
10. Chute extensions to be as near vertical as possible but not less than 50 per cent angle of inclination to the horizontal.
11. Extensions can be bifurcated to allow discharge left or right.

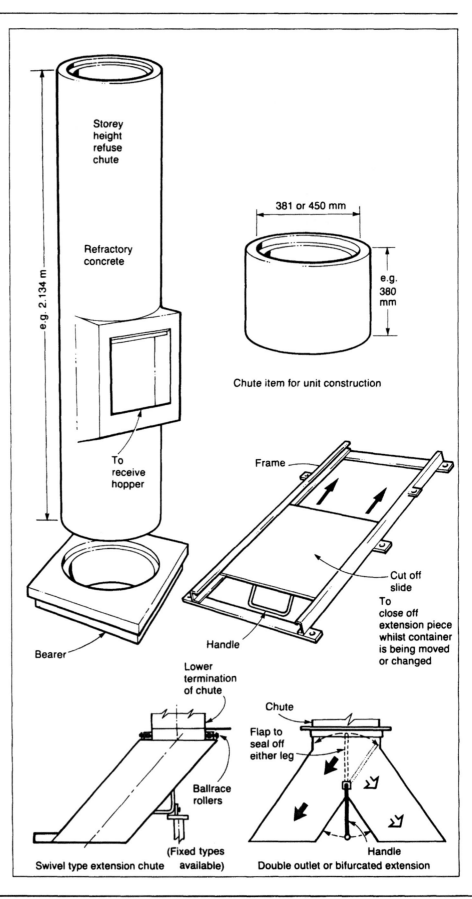

Storey height refuse chute

Refractory concrete

e.g. 2.134 m

To receive hopper

Bearer

381 or 450 mm

e.g. 380 mm

Chute item for unit construction

Frame

Cut off slide

To close off extension piece whilst container is being moved or changed

Handle

Lower termination of chute

Ballrace rollers

(Fixed types available)

Swivel type extension chute

Chute

Flap to seal off either leg

Handle

Double outlet or bifurcated extension

Sanitation: Refuse From Buildings

DESIGN NOTES. HOPPERS

1. Whenever the design of the building permits, hoppers should be in freely ventilated position in open air e.g. sheltered balconies.
2. They should not be situated within any stairway enclosure, enclosed staircase lobby or enclosed corridor but, where a hopper is approached from such enclosed circulation routes, the chute and hopper should be contained within its own compartment constructed of fire-resisting material; self-closing access door and the compartment freely ventilated to the external air.
3. Adequate ventilation of such a lobby is of paramount importance (minimum a/c per hour) and maximum advantage of daylight be taken, plus the provision of artificial lighting.
4. In no case should hoppers be situated within a dwelling or habitable room or place used in connection with preparation of food.
5. Hoppers should be fixed at a height of 750 mm measured from floor level to the lower edge of the inlet opening.
6. Maximum size of clear opening of hopper mouth — 250 mm high and 350 mm wide; to be designed to prevent emission of dust or fumes; to prevent noise in use; and cause no obstruction to free passage or refuse.
7. Problems of blockages have made it necessary to limit the size of the hopper opening in relation to the size of chute.
8. Wall and floor construction to be smooth, impervious and easy to clean.

VENTS

1. Cross section of vent and total area of outlet to be minimum 150 mm or 10 per cent of the cross-section of area of chute, whichever is greater, but full bore continuation of chute to external air is recommended.
2. Vent pipe to be protected at its outlet against weather and birds; be as near vertical as possible; offsets minimum 45 deg. horizontal and provided with access. Common vents can be used.

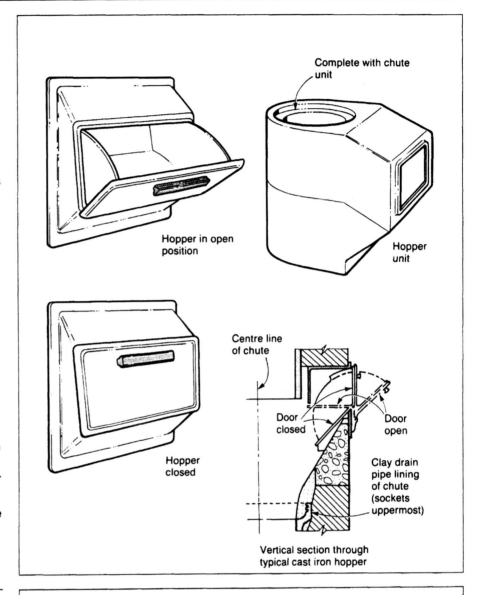

Hopper in open position

Complete with chute unit

Hopper unit

Hopper closed

Centre line of chute

Door closed

Door open

Clay drain pipe lining of chute (sockets uppermost)

Vertical section through typical cast iron hopper

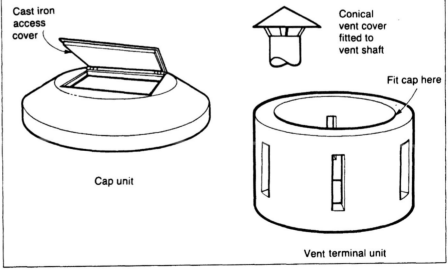

Cast iron access cover

Cap unit

Conical vent cover fitted to vent shaft

Fit cap here

Vent terminal unit

Dry Chute Systems

ADVANTAGES

1. Convenient to use if chutes are properly located.
2. Low capital cost, particularly if the number of chutes is reduced to the minimum, e.g. 1 chute serving many containers.
3. Low cost of collection without inconvenience to the householders.

DISADVANTAGES

1. Possibility of smells and fumes if chute not adequately ventilated or badly positioned.
2. Large items not capable of passing down the chute (or through the hopper) must be carried to ground level.
3. A chute with an I.D. of less than 375 mm is liable to blockage. The trend is towards chutes of 450 mm internal diameter.
4. There may be a noise problem if the hoppers are not properly designed and correctly positioned.
5. The disposal of hot ashes may create a fire hazard with the attendance risk of a serious smoke nuisance.

TYPICAL CALCULATION

Assume 8 storey block of flats with 4 flats per floor and 4 people per flat. Allow for one collection per week and the use of British Standard cylindrical receptacles.

Volume to be expected:
$8 \times 4 \times 4 \times 0.03 \text{ m}^3 = 3.84 \text{ m}^3$
(Note: 0.03 m³/person per week used).

Total storage capacity (allowing for one week's accumulation) is therefore 3.84 m³.

Capacity of one receptacle (cyl. type) = 0.95 m³.

Therefore number of receptacles =
$\frac{3.84}{0.95} = \text{say 4}.$

Number of receptacles per chute — assume 2.

Therefore two chutes are required with access to each chute on alternative floors. This precaution will ensure even loading of the receptacles.

Bifurcation is necessary at the bottom of each chute.

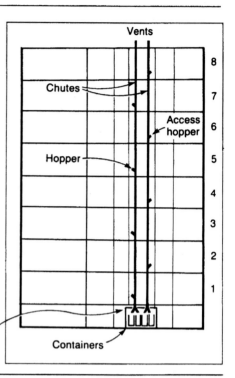

CHOICE OF METHOD

Houses and Bungalows. Storage and collection of refuse is best effected by the provision of individual storage containers for each dwelling but in some circumstances, for groups of houses, communal containers could be used.

Dwellings in low blocks (three storeys in flats and four storeys in maisonettes). Refuse storage containers for communal use with chutes where practicable are recommended for this type of development. The chutes should be spaced at not more than 60 m intervals on the assumption that an occupier should not be required to carry refuse more than 30 m.

Dwellings in higher blocks. Communal chutes are recommended for dwellings in higher blocks.

Sanitation: Refuse From Buildings

ROTARY BINS SYSTEMS

The 'Binrota' is a rotary turret holding 3 to 12 bins of up to 0.57 or 0.85 m³ capacity but can be adapted to fit any type and size required. An empty bin is manually rotated to its position beneath a refuse chute, precise location by action of spring-loaded locking pin. As one bin is filled, the next one can be rotated into position.

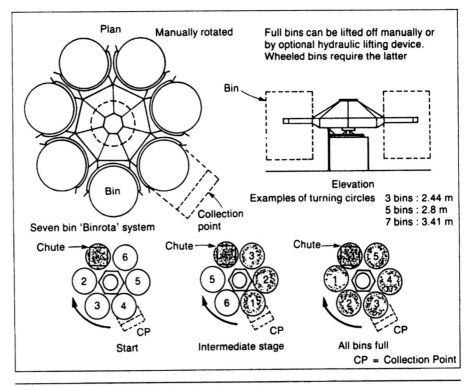

Plan Manually rotated

Full bins can be lifted off manually or by optional hydraulic lifting device. Wheeled bins require the latter

Bin

Elevation

Examples of turning circles

3 bins : 2.44 m
5 bins : 2.8 m
7 bins : 3.41 m

Seven bin 'Binrota' system

Start

Intermediate stage

All bins full

CP = Collection Point

BULKY HOUSEHOLD DISCARDS

The Civic Amenities Act 1967 lays on Local Authority's the duty of providing places where refuse, other than business refuse, may be deposited by local residents free of charge, and by other persons on payment of such charge as the Local Authority think fit. Sites may be open for the deposit of refuse at all reasonable times. For special collection service of bulky or other household discards, refer to Local Authority.

ACCESS FOR VEHICLES

Roadways:- to have foundations and hard wearing surface to withstand the loading imposed by collecting vehicles. Minimum width:- 5.0 m and arranged so that vehicles can continue in forward direction. Turning circles should be related to the largest vehicle — consult local authority. Paths between container chamber and vehicle to be free from kerbs; minimum 2 m wide; level or to slope to vehicle; maximum gradient 1 in 12.

SACK COMPACTION SYSTEMS

This is the Opima Carousel refuse compactor from C. S. Bacon and Son (London), It is suitable for serving from 8 to 40 dwellings, and is available in four capacities with either four, six, eight or ten sacks.

This system uses a power-operated ram to compress refuse into disposable sacks. Up to 10 sacks are generally carried on a turntable installed in a chamber under each chute. The compaction cycle is started when refuse falling down the chute operates a sensitive device which operates a cut-off plate and the refuse is forced into the sack by a ram. When the refuse reaches a pre-determined level, after repeated filling and compression, the turntable rotates to bring an empty sack to the filling position. Compaction ratios up to 4:1 can be attained with normal refuse but 8:1 is possible, and even 15:1 has been recently introduced.

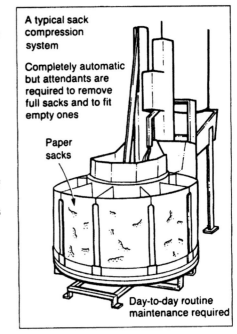

A typical sack compression system

Completely automatic but attendants are required to remove full sacks and to fit empty ones

Paper sacks

Day-to-day routine maintenance required

Pneumatic Conveying Systems

DIAGRAMMATIC ARRANGEMENT OF VALVE ROOMS AND AIR INLET

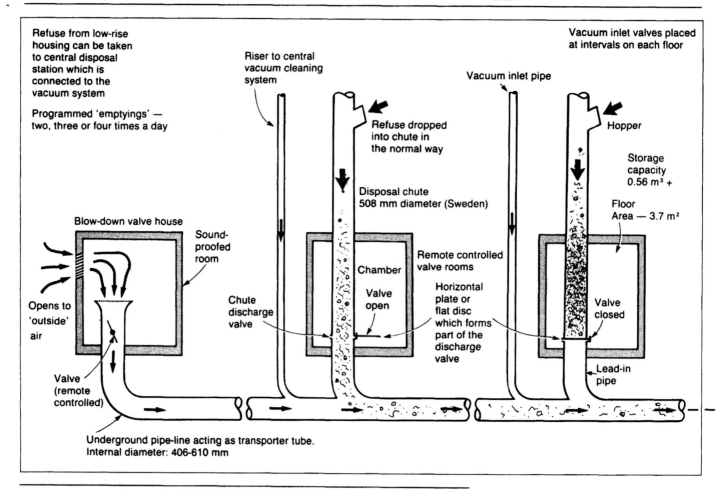

Refuse from low-rise housing can be taken to central disposal station which is connected to the vacuum system

Programmed 'emptyings' — two, three or four times a day

Vacuum inlet valves placed at intervals on each floor

Riser to central vacuum cleaning system

Vacuum inlet pipe

Refuse dropped into chute in the normal way

Hopper

Disposal chute 508 mm diameter (Sweden)

Storage capacity 0.56 m³ +

Blow-down valve house

Sound-proofed room

Floor Area — 3.7 m²

Remote controlled valve rooms

Chamber

Opens to 'outside' air

Chute discharge valve

Valve open

Horizontal plate or flat disc which forms part of the discharge valve

Valve closed

Valve (remote controlled)

Lead-in pipe

Underground pipe-line acting as transporter tube. Internal diameter: 406-610 mm

OPERATION

This system requires the use of conventional refuse chutes or, alternatively, where it is accepted that tenants may carry their refuse to ground floor level, then the use of chutes connected to a central loading point. It can handle any refuse which can normally be put in the chute. The lower end of the chute terminates in a valve chamber, where the chute is connected by lead-in pipe to horizontal steel transporter pipe(s) of approx. 525 mm diameter which are laid below ground. Refuse falling down the chute rests on a flat valve disc at the bottom of the chute.

This valve connects with the lead-in pipe. The functions of the valve are to provide a base upon which refuse in the chute can be stored and to create a seal when vacuum is being created in the transporter pipes by extractor turbines. The valves in the system are opened at predetermined intervals and the refuse is 'dumped' into the tube and carried along the underground pipeline at high speed to a central collecting bin. Starting and switch operations are executed at intervals and it is possible to clear 144 chutes in 30 minutes:- average 12 sec. per chute.

Sanitation: Refuse From Buildings

DIAGRAMMATIC ARRANGEMENT OF PLANT ROOM

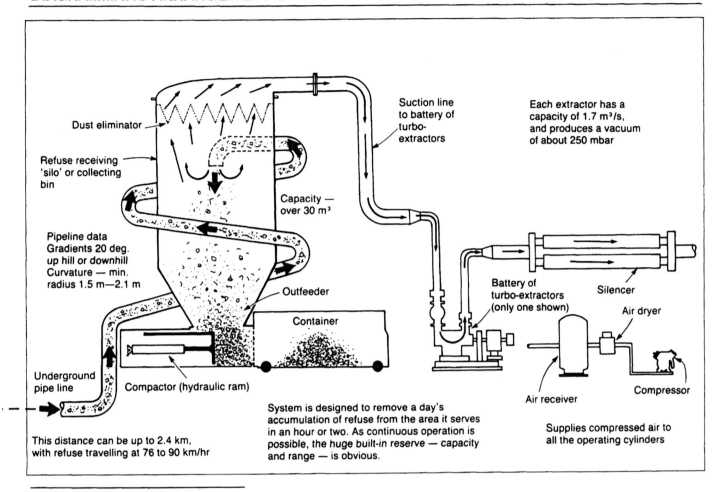

Dust eliminator

Refuse receiving 'silo' or collecting bin

Suction line to battery of turbo-extractors

Each extractor has a capacity of 1.7 m³/s, and produces a vacuum of about 250 mbar

Capacity — over 30 m³

Pipeline data
Gradients 20 deg. up hill or downhill
Curvature — min. radius 1.5 m—2.1 m

Outfeeder

Container

Battery of turbo-extractors (only one shown)

Silencer

Air dryer

Underground pipe line

Compactor (hydraulic ram)

Compressor

Air receiver

This distance can be up to 2.4 km, with refuse travelling at 76 to 90 km/hr

System is designed to remove a day's accumulation of refuse from the area it serves in an hour or two. As continuous operation is possible, the huge built-in reserve — capacity and range — is obvious.

Supplies compressed air to all the operating cylinders

EVACUATION CYCLE

At the scheduled evacuation times, a group of turbo pumps at the central refuse collection point start running to depressurise the tube network. A valve opens to outside air; the chute valves open one at a time, and the refuse is sucked to the collection point, compressed into containers and hauled away. Spent air escapes from the 'silo' by passing through high-efficiency dust filters, then vented through silencers to atmosphere. Refuse collects at base of 'silo' and is rammed into special containers for disposal. The entire process is automatically run according to a programme 'made-to-measure' for each particular installation. Vacuum conveyance facilitates the handling of dirty laundry, via a separate tube system, by vacuum tapped from the same power source. Incineration on site, discharge to hopper, or combined with hammer mills for homogenisation are alternatives for *refuse disposal.*

Printed in the United Kingdom
by Lightning Source UK Ltd.
125051UK00010B/15-22/A